DIPLOMACY UNPLUGGED

REIMAGINING GLOBAL ENGAGEMENT IN A DIGITAL WORLD

A Comprehensive Exploration of Technology's Impact on 21st Century Diplomacy

DR. OLEG FIRER

Copyright © 2025 Dr. Oleg Firer. All rights reserved.

No part of this book may be reproduced, stored in a retrieval system, or transmitted in any form or by any means without the written permission of the author.

First Edition

ISBN 979-8-9913195-2-2

"Diplomacy is no longer confined to closed chambers or whispered corridors. It lives in the pulse of platforms, the flow of code, and the voices of people connecting across borders."

PREFACE

The pace of change in our global society is relentless. Each new sunrise seems to bring with it another innovation, another platform, another disruption that redefines how individuals, institutions, and entire nations communicate and collaborate. In this new world, diplomacy has not been spared—it, too, must evolve.

For centuries, diplomacy was conducted in private rooms, shaped by ritual, guided by protocols, and influenced by individuals who had spent decades mastering the art of discretion. It was a world governed by long timelines and deliberate silences. But that world, while still echoing in many institutions, no longer defines the present. In its place stands a more unpredictable, more transparent, and more participatory form of global engagement—one mediated by screens, shaped by algorithms, and played out in real time before a global audience.

This book was born out of a conviction that we are at a turning point in international relations—a moment when the old tools are no longer sufficient, and the new ones are still being understood. I have lived and worked at this intersection, balancing tradition with innovation, formality with digital fluidity. From closed-door negotiations to viral hashtags, from carefully worded communiqués to spontaneous livestreams, I have witnessed diplomacy being rewritten, not with ink, but with code.

"Diplomacy Unplugged" is not intended to replace traditional models of diplomacy. Rather, it aims to supplement them—to reflect on how the digital age has disrupted our assumptions, challenged our institutions, and introduced new actors, new platforms, and new risks into the diplomatic arena. It is a book grounded in real-world observation but driven by a deep belief that diplomacy, if it is to survive, must become as dynamic and responsive as the world it seeks to serve.

To the reader—whether you are a policy advisor, a student of international affairs, a technologist shaping the next platform, or a global citizen interested in how nations navigate conflict and cooperation—this book is for you. It is a call to engage with the challenges ahead, not with fear, but with clarity, intention, and resolve.

ACKNOWLEDGEMENTS

No book—especially one that explores the intersections of diplomacy, technology, and global transformation—is created in isolation. While the digital age may accelerate the pace of writing and research, the true heart of this project lies in human connection: in long conversations, quiet mentorship, and the courage of those shaping our shared future one decision at a time.

First and foremost, I offer my deepest gratitude to the diplomats, policy leaders, technologists, academics, and activists who opened their time, stories, and perspectives to me. Many of you spoke candidly about your fears, your breakthroughs, and your hope for a more just and humane international system. These conversations—sometimes in boardrooms, sometimes over coffee, and often across screens—shaped the living foundation of this book.

To the researchers, analysts, and subject matter experts who helped bring texture and accuracy to the chapters ahead: your scholarship gave structure to insight and kept this work grounded in reality. It is one thing to explore bold ideas—it is another to do so with fidelity to fact and discipline.

To the silent architects of diplomacy—the individuals whose names don't appear in headlines, whose work is done behind secure doors or encrypted lines—thank you. Your quiet courage and professionalism sustain global peace in ways that too often go unrecognized.

To my colleagues in multilateral institutions, national ministries, civil society organizations, and international think tanks: your tireless efforts to adapt diplomacy to our digital era inspire confidence in the future of global cooperation. This book is enriched by the many partnerships, dialogues, and debates we've shared over the years.

To the editors and early readers who pushed me to clarify, to question, and to dig deeper: you made this book better. You reminded me that clarity is a form of respect, and that thoughtful writing begins with thoughtful listening.

And most personally, to my family—your unwavering patience, encouragement, and love have been my constant source of strength. To my daughter Evette, whose boundless curiosity and incisive geopolitical questions serve as a daily reminder that diplomacy is not limited to official corridors—it lives in the inquisitive minds of the next generation, across dinner tables, and within every thoughtful "why." You inspire me to remain ever curious, ever engaged. To my daughter Mirella, whose presence brings balance, light, and joy to every chapter of my life. To my wife, Karina, my steadfast partner, thank you for standing beside me through every late night, every journey, and every aspiration. And to my mother, Larisa, whose quiet and unwavering faith in me has been the foundation of my resilience—thank you for believing in this vision long before it took form.

And to the reader—thank you. Whether you come to these pages as a diplomat, a student, a skeptic, or a dreamer, I am honored to be in conversation with you. May this book spark new insights and remind us all that diplomacy, at its core, is a human endeavor—fragile, complex, and full of potential.

— *Oleg Firer*

CONTENTS

PREFACE 2
ACKNOWLEDGEMENTS 4
INTRODUCTION 9
PART I: FOUNDATION OF MODERN DIPLOMACY 15
CHAPTER 1: THE EVOLUTION OF DIPLOMACY 16
CHAPTER 2: DIGITAL DISRUPTION IN STATECRAFT 26
CHAPTER 3: SOCIAL MEDIA AS A DIPLOMATIC TOOL 36
CHAPTER 4: WAR WITHOUT BORDERS 61
CHAPTER 5: INFORMATION WARFARE: 90
CHAPTER 6: THE ROLE OF AI IN DIPLOMACY 104
PART II: NARRATIVE AND PERCEPTION 113
CHAPTER 7: CRISIS DIPLOMACY IN THE DIGITAL AGE ...114
CHAPTER 8: CYBERSECURITY CHALLENGES 144
CHAPTER 9: CRISIS MANAGEMENT AND COMMUNICATION IN THE DIGITAL ERA 153
CHAPTER 10: THE GEOPOLITICS OF DATA AND SOVEREIGNTY 177
CHAPTER 11: THE EVOLVING ROLE OF NON-STATE ACTORS IN DIGITAL DIPLOMACY 201
PART III: IDENTITY, RIGHTS & INCLUSION 221
CHAPTER 12: GENDER AND DIGITAL DIPLOMACY 222
CHAPTER 13: THE ROLE OF CULTURE IN DIPLOMACY ...237
CHAPTER 14: DIGITAL IDENTITY, CITIZENSHIP, AND INCLUSION 252

Chapter 15: The Impact of Technology on Human Rights Advocacy..266
Chapter 16: Digital Ethics and Responsibilities ..281
Chapter 17: The Battle for Truth..............................295
Part IV: Digital Power and Sovereignty 310
Chapter 18: Blockchain and Decentralized Governance..311
Chapter 19: The Role of Digital Infrastructure in Development and Sovereignty...................................324
Chapter 20: Algorithms at the Table........................336
Chapter 21: Cooperation in the Digital Age 350
Chapter 22: The Future of Global Governance in the Digital Age..364
Part V: The Future of Diplomacy 374
Chapter 23: Education and Training for Digital Diplomacy 375
Chapter 24: Recommendations for Practitioners and Policymakers..393
Chapter 25: Reflections on Digital Diplomacy: Embracing Change for a Secure Future 402
Chapter 26: Conclusion 411
Chapter 27: Epilogue..416
Glossary of Digital Diplomacy Concepts 418
Bibliography..423

INTRODUCTION

DIPLOMACY'S NEW ARENA: REIMAGINING GLOBAL ENGAGEMENT IN A CONNECTED AGE

There was once a rhythm to diplomacy that felt deliberate, dignified, and deeply human. It unfolded in carefully orchestrated stages—in handwritten notes passed through trusted envoys, in discreet conversations behind closed doors, in communiqués drafted with surgical precision and sometimes never released at all. The stakes were high, but the process was paced. Diplomacy was an art of balance—of words weighed not only for what they said, but for what they left unsaid. Silence was often more strategic than speech, and the ability to withhold a reaction was sometimes the most powerful move of all. A crisis could be contained before it ever reached the public eye, and a misunderstanding could be corrected before it hardened into consequence.

The first time I participated in an international negotiation, I understood this rhythm intuitively. The room was modest, the setting unremarkable, but the air was charged with an invisible tension that had less to do with the words being spoken and more with the decades—sometimes centuries—of history embedded within each phrase, each pause, each calculated gesture. It was a dance of nuance. The outcome mattered, of course, but so did the process—the choreography of mutual recognition, the delicate symmetry of power and restraint.

That world, however, no longer exists in the form we once knew. Today, diplomacy operates in an entirely different arena—one shaped not by polished wood tables and quiet deliberation, but by screens, speed, and the relentless churn of global visibility. It no longer begins and ends in foreign ministries; it unfolds across social media feeds, livestreamed press conferences, digital leaks, and online campaigns. A single tweet—well-timed or poorly worded—can provoke international backlash. A manipulated video can upend years of relationship-building. Conversations once held in confidentiality now occur in public, sometimes involuntarily, as the line between internal dialogue and global broadcast continues to dissolve.

What has emerged is not simply a faster version of diplomacy, but an entirely new mode of global engagement—one in which perception often moves faster than policy, influence is dispersed across unexpected actors, and the boundaries of diplomatic space are constantly shifting. Governments are no longer the sole stewards of foreign policy. Platforms, corporations, influencers, activists, and algorithms have joined the fray—some intentionally, others by default. The ability to shape global narratives, command public attention, and negotiate power dynamics no longer belongs exclusively to those with diplomatic passports or formal mandates.

This shift, while disruptive, is not inherently destructive. It represents a profound transformation—one that carries risks, but also extraordinary possibilities. The architecture of diplomacy is being rebuilt in real time, and with it, the very meaning of participation in international affairs is being redefined. The challenge we face is not simply to keep up, but to guide that transformation with intention, ethics, and foresight.

This book was written in response to that challenge—and out of a deep conviction that diplomacy remains not only relevant, but vital. In a world saturated with noise, polarization, and disinformation, we need more than ever the practices that diplomacy was built upon deep listening, careful framing, strategic patience, and principled engagement. But we must also acknowledge that the methods, spaces, and actors involved in this work have changed—and will continue to change. What is needed now is not nostalgia, but adaptation. Not preservation of form, but renewal of purpose.

To that end, this book is neither a how-to guide for tweeting like a diplomat nor a retrospective on lost traditions. It is a human-centered exploration of diplomacy in the digital age—an inquiry into how trust, legitimacy, influence, and ethics are negotiated across a hyperconnected world. It is grounded in personal experience across government, technology, and policy arenas, but it is written for a much broader audience—because in the 21st century, diplomacy no longer belongs only to diplomats.

HOW THE BOOK IS STRUCTURED

To help navigate this complex landscape, the book unfolds in five thematic parts. Each section builds on the last, creating a cumulative lens through which to understand the changing shape of diplomacy.

Part I: Foundations of Modern Diplomacy

We begin by tracing the roots of diplomacy, examining how long-standing values like discretion, sovereignty, and legitimacy are being challenged—and sometimes redefined—by digital realities. We explore how education, credibility, and institutional norms must evolve alongside our tools.

Part II: Narrative and Perception

This section delves into the new frontlines of influence—where perception often outpaces policy. From disinformation campaigns to the algorithmic war for attention, we look at how reputation, emotion, and narrative are now central to international strategy.

Part III: Identity, Rights & Inclusion

Diplomacy is not only about power, but also about representation. Here, we explore the ethical tensions of online engagement—who gets heard, who is excluded, and how identity, gender, and culture are negotiated and contested in the digital arena.

Part IV: Digital Power and Sovereignty

Diplomacy depends on systems we can no longer see but increasingly rely on. This section addresses the physical and political infrastructure behind global engagement—from internet governance and data sovereignty to platform accountability and digital inclusion.

Part V: The Future of Diplomacy

We conclude by considering the future: What skills will tomorrow's diplomats need? How must institutions evolve? What ethical and inclusive global engagement can help rebuild trust in a fractured world?

This book isn't a definitive answer but offers a framework for understanding change and encourages constructive engagement. I believe that true diplomacy still has the power to unite us, even amid chaos.

Let's begin with curiosity, resolve, and togetherness.

Part I: Foundation of Modern Diplomacy

Chapter 1: The Evolution of Diplomacy

FROM SEALS TO SCREENS

From wax seals impressed on clay tablets in ancient kingdoms to handwritten scrolls exchanged by royal couriers, the art of diplomacy began as a slow, deliberate process. It was reserved for sovereigns and their trusted emissaries—ritualistic, secretive, and often life-or-death in its stakes. Yet, even then, the essential purpose was recognizable: to communicate across distance, resolve disputes without violence, and protect shared interests.

Over centuries, each technological leap—from writing systems to digital networks—transformed how those aims were pursued. Where once messages took weeks by horseback, today's negotiations unfold in seconds across encrypted video links or real-time social media threads. But while the methods have changed dramatically, the heart of diplomacy remains the same:

the careful, strategic management of relationships between peoples and powers.

This chapter traces that transformation. It's a journey from the formal and ceremonial diplomacy of empires to today's fast-paced, decentralized, and hyper-public practice. And it offers a reminder: if diplomacy has always evolved alongside communication tools, then today's digital shift—disruptive though it may be—is part of a long, human story.

ANCIENT AND MEDIEVAL DIPLOMACY

In ancient civilizations—from Egypt and Mesopotamia to Persia, India, and China—leaders exchanged messages carved into stone or written on papyrus. These messages often carried news of alliances, trade offers, or strategic marriages. Diplomats were trusted envoys, walking or sailing great distances to represent the voice of their king or ruler.

By the medieval period, diplomacy had begun to take on a more structured form. In Europe, permanent embassies started appearing in key cities, particularly in the Italian states. The role of an envoy became a refined position—one bound by protocol, etiquette, and codes of secrecy. One misstep in tone or gesture could shift the direction of a negotiation, or worse, provoke a conflict.

Diplomatic immunity—a concept that today feels standard—emerged from this period as a recognition of how vital, and vulnerable, these early diplomats were. Harming an envoy was seen as not just a crime, but a provocation of war.

These traditions laid the groundwork for what would become modern diplomacy: formal, confidential, and carefully choreographed.

FROM EMPIRE TO EMBASSY

In antiquity, diplomacy was a high-risk endeavor conducted under strict codes of honor and protocol. Envoys were often seen as sacred messengers, protected by custom, though not always by force. They carried symbols of peace, clay tablets or scrolls inscribed with the marks of rulers—seals that conveyed authenticity in a world where the spoken word could vanish with the wind.

These early diplomatic exchanges were not simply informational—they were performative, theatrical, and deeply embedded in cultural ritual. The form mattered as much as the content.

THE RISE OF WRITTEN DIPLOMACY

The invention of writing revolutionized diplomacy. Agreements could now be recorded, referenced, and transmitted more reliably. It wasn't just about what was said, but what could be proved.

This gave rise to protocols, treaties, and the notion of a diplomatic "record." States began to keep archives. Envoys carried letters of credence. Formal rules developed—many of which still echo in today's Vienna Convention on Diplomatic Relations.

With writing came continuity. Successive rulers could reference past agreements. Complex negotiations no longer depended on oral memory. And diplomacy began its transformation from performance to profession.

THE PRINTING PRESS AND THE PUBLIC SPHERE

The 15th-century printing press shattered the exclusivity of knowledge. Suddenly, ideas could travel faster and farther than any diplomat. Pamphlets, treatises, and political commentaries began circulating among an increasingly literate public.

This changed diplomacy in two profound ways. First, it made governments more accountable. Citizens could now form opinions about foreign policy. Second, it introduced the concept of public diplomacy—where states not only negotiated with each other, but also sought to shape the hearts and minds of foreign populations.

Diplomacy was no longer confined to royal courts. It entered the public square.

THE TELEGRAPH AND THE COLLAPSE OF DISTANCE

The 19th century marked the beginning of real-time diplomacy. The telegraph allowed capitals to communicate instantly with far-flung embassies. Crises could be managed in hours instead of weeks.

But this newfound speed brought tension. Ambassadors—once powerful decision-makers—found their autonomy reduced. Instructions now flowed from capitals. Diplomacy became more centralized, more reactive, and more entwined with domestic politics.

The telephone continued this trend. Voice replaced parchment. Urgency became a constant.

BROADCAST DIPLOMACY: SPEAKING TO THE WORLD

The 20th century brought radio and television into the diplomatic toolkit. Leaders could now speak directly to foreign populations—bypassing traditional diplomatic channels.

Franklin Roosevelt's fireside chats, Winston Churchill's wartime broadcasts, and John F. Kennedy's Cold War speeches weren't just moments in media history—they were acts of statecraft.

Mass communication added a new layer to diplomacy: performance. Leaders became storytellers. Messages were crafted for both allies and adversaries. Optics mattered. So did tone, emotion, and timing.

The diplomat's role shifted again—from negotiator to communicator.

THE INTERNET: A NEW ARENA, NOT JUST A NEW TOOL

If the printing press democratized knowledge, the internet democratized power.

Suddenly, anyone could publish. Anyone could respond. Information moved at the speed of light—and so did misinformation.

Traditional diplomacy, built on confidentiality and slow deliberation, struggled to keep up. Leaks became commonplace. Whistleblowers and hackers became geopolitical actors. Activists and influencers could shape global narratives from a smartphone.

Diplomacy was no longer a conversation between states. It became a dynamic, decentralized, and often unpredictable ecosystem.

SOCIAL MEDIA AND THE PERSONALITY OF POWER

The rise of social media marked a turning point. World leaders created personal Twitter accounts. Foreign ministries live-streamed negotiations. A viral video could shift public sentiment faster than a press release.

This created opportunities—for transparency, immediacy, and direct engagement. But it also carried risks. Diplomatic nuance was often lost in 280 characters. Misinformation spread rapidly. Emotional outbursts became geopolitical events.

Diplomacy, once the art of quiet influence, became increasingly performative.

DIPLOMACY IN THE AGE OF ALGORITHMS

Today, diplomacy operates within digital ecosystems that are shaped by algorithms, data analytics, and artificial intelligence. Attention is the new currency. Narrative control is power.

A single tweet can influence markets. A trending hashtag can pressure governments. A well-placed leak can derail years of negotiation.

Diplomats now work alongside data scientists, cybersecurity teams, and communications strategists. Influence is measured not only in alliances but in engagement metrics. Trust is built—and eroded—in public view.

This is diplomacy in the age of code.

CONCLUSION: A LEGACY REWRITTEN IN CODE

As we look back across millennia—from clay tablets to encrypted chats—it becomes clear that diplomacy has never stood still. It has responded, adapted, and often led the way in redefining how societies communicate and cooperate.

Each communication revolution has done more than alter speed or style; it has reshaped who gets to speak, who gets to be heard, and how legitimacy is conferred. In today's world, digital transformation is not merely an overlay—it is a rewrite of the diplomatic script.

Yet even amid disruption, the essence remains. Diplomacy is still about navigating complexity with care. It is still about forging connection where there is distance, building trust where there is suspicion, and seeking peace where there is tension.

The tools may have changed. The settings may be virtual. But the stakes remain deeply human.

As we step into the next chapter, we shift from a historical view to a real-time reality. We'll explore how modern technology—from blockchain to artificial intelligence—is actively reshaping diplomatic processes, redefining borders, and recalibrating influence. Diplomacy is no longer just about words—it's about code. And that code, increasingly, governs our world.

Chapter 2: Digital Disruption in Statecraft

WHEN CODE BECOMES CURRENCY

There was a time—not so long ago—when diplomats could control the timing, tone, and tools of international engagement. A carefully worded statement could be polished overnight. A policy shift could be announced with orchestrated precision. Negotiations, often cloaked in discretion, were released to the public only when the moment was right.

That era has ended.

In its place stands a new diplomatic landscape: raw, rapid, and relentlessly digital. A livestreamed press conference, a viral tweet, or a cyber incident can shape international narratives faster than any formal communiqué. We are no longer simply using technology in diplomacy. We are operating within it. Technology is no longer the instrument—it is the stage, the terrain, and increasingly, the actor.

This chapter explores what it means to practice diplomacy in a world shaped by algorithms, networks, and platforms. It is about how states navigate a global environment where influence is quantified in likes and shares, where negotiation takes place over webcams, and where the most important geopolitical shifts may begin not in embassies, but in data centers.

REDEFINING DIGITAL DIPLOMACY

Let's begin with clarity. "Digital diplomacy" is more than a buzzword or a social media campaign. At its core, it refers to the strategic use of digital technologies by governments, diplomats, and international institutions to achieve foreign policy goals. That includes tools like social media, encrypted communication platforms, AI-driven analysis, digital currencies, and virtual conferencing.

But that's just the beginning.

Digital diplomacy represents a broader transformation in how statecraft is practiced. It reflects a shift in mindset—from closed-door deliberation to participatory influence, from top-down messaging to adaptive engagement, and from physical stages to fast-moving, volatile digital ecosystems. In this new world, reputation, responsiveness, and narrative framing are not just helpful—they are essential to diplomatic credibility.

Digital diplomacy now encompasses multiple, overlapping domains:

- **Public diplomacy** through social media, real-time video messaging, and interactive campaigns.
- **Virtual diplomacy**, where high-level negotiations occur through platforms like Zoom and Teams.
- **Policy enablement**, such as digital identity systems, blockchain-driven agreements, and fintech integration into state infrastructure.
- **Strategic digital influence**, through sentiment analysis, narrative shaping, and digital listening tools.
- **Platform diplomacy**, where diplomats engage not just with other governments, but with tech companies and infrastructure providers—new power brokers in a borderless digital world.

These are not separate channels. They are converging forces. Together, they are not just reshaping diplomacy—they are redefining the very context in which it operates.

THE TOOLS OF THE NEW DIPLOMACY

The diplomat of the 21st century needs more than briefing notes and diplomatic cables. They need data dashboards, real-time monitoring systems, a strong grasp of digital culture, and the ability to interpret trends faster than they unfold.

Let's look at the core technologies reshaping how diplomacy is done:

1. Social Media and Direct Engagement

Platforms like X (formerly Twitter), Facebook, LinkedIn, TikTok, and YouTube are now frontline diplomatic tools. A single tweet from a foreign minister can offer reassurance—or spark a geopolitical firestorm. Embassies livestream updates. Heads of state use personal accounts to shape narratives directly, bypassing the filter of traditional media.

Influence is now measured in impressions, shares, and engagement—not just in paragraphs or protocol. Diplomats must understand how digital tone, timing, and cultural fluency affect credibility on the global stage.

2. Virtual Diplomacy and Remote Summits

The COVID-19 pandemic accelerated a transformation already underway. Today, diplomacy increasingly takes place on screen. Summits are hosted over Webex. Multilateral consultations occur on Zoom. Crisis talks unfold across secure channels in real time.

While this has increased accessibility and speed, it also alters the subtle mechanics of negotiation. Physical presence, body language, and informal sideline conversations—the soft power of traditional diplomacy—are harder to replicate in digital spaces.

3. Fintech and Digital Currency Frameworks

As digital currencies and fintech tools become central to national economic strategies, diplomacy must evolve accordingly. Whether negotiating central bank digital currency (CBDC) interoperability, remittance frameworks, or smart contract regulation, diplomats now need to understand the technical underpinnings of digital finance.

Global influence is increasingly exercised through APIs, regulatory sandboxes, and blockchain governance discussions—not just through bilateral bank accords.

4. Big Data and Sentiment Intelligence

Diplomatic success requires situational awareness. Digital diplomacy is increasingly powered by real-time data analytics, which track global sentiment, flag misinformation trends, and assess potential unrest or escalation.

Nations now use social listening tools to anticipate reputational risks and preempt crises. Digital intelligence doesn't replace traditional diplomacy—it augments it with actionable insight.

5. Cloud Platforms & Cross-Border Collaboration

From climate agreements to humanitarian coordination, cloud-based collaboration platforms now enable treaty drafting, interagency cooperation, and multilateral project oversight across borders. These tools bring speed, transparency, and inclusivity—but also introduce challenges in cybersecurity, data sovereignty, and interoperability.

DISRUPTION IN ACTION: WHAT'S CHANGING

We are witnessing more than a technological shift—we are navigating a full paradigm change. Here's how the practice of diplomacy is being disrupted in real time:

1. Speed and Agility Replace Bureaucracy

Crises now erupt—and escalate—in minutes. A single social media post can trigger public outcry, shift market sentiment, or pressure a government into action. In this environment, long decision chains and cautious positioning may be liabilities. Diplomats must be agile, responsive, and digitally fluent.

2. Non-State Actors Move Center Stage

Influencers, tech executives, NGOs, and cyber activists increasingly shape international outcomes. A tweet from a CEO may trigger more reaction than a communique from a foreign ministry. A viral campaign can change the terms of a trade deal.

Diplomacy is no longer a state-exclusive domain. It now plays out in a crowded arena, where legitimacy and authority must be earned across sectors.

3. Power Is Fragmented and Privatized

Many critical levers of the digital world are held by private companies—cloud providers, platform owners, telecom giants. This creates new diplomatic dynamics, where states must negotiate with entities they do not control, often in opaque or asymmetric contexts.

4. Crises Are Public and Real-Time

Gone are the days when a crisis could be managed quietly. Today, if a consulate is attacked, images and videos circulate before embassies release a statement. The window to shape narratives is measured in minutes, not hours. The difference between perception and reality widens fast.

5. Trust Must Be Rebuilt—Constantly

In an era of deepfakes, misinformation, and surveillance anxiety, trust in institutions is fragile. Authority no longer flows from the title "Ambassador" or "Minister." It flows from consistent transparency, credibility under pressure, and the ability to humanize diplomacy in real time.

THE ETHICAL CROSSROADS

With new tools come new responsibilities—and moral dilemmas. Should governments run targeted influence campaigns abroad? Should diplomats engage with platforms that suppress dissent or spread propaganda? What happens when surveillance technologies are used in the name of security, but at the cost of democratic freedoms?

In a world where platforms shape perceptions and algorithms influence outcomes, who is accountable when diplomacy fails—or escalates?

Ethical leadership is not optional. It is the defining test of digital diplomacy.

CONCLUSION: DIPLOMACY IN THE DISRUPTED ERA

The digital disruption of diplomacy is not a future scenario. It is today's reality.

It has unsettled long-standing norms, challenged the role of the state, and forced diplomats into unfamiliar terrain. But it also offers immense opportunity—for more inclusive engagement, faster crisis response, and reimagined forms of global cooperation.

Diplomacy is not dying—it is digitizing.

The next chapter explores the frontlines of this transformation: how social media has become both a megaphone and a minefield for diplomats, and why mastering its use is now central to any global engagement strategy.

Chapter 3: Social Media as a Diplomatic Tool

WHERE DIPLOMACY MEETS THE SCROLL

Once upon a time, diplomacy moved at the pace of dispatches and couriers. Messages took days to arrive, and each word passed through layers of refinement before reaching the public. That era is over. Today, the scroll of a smartphone has replaced the slow turn of parchment. Social media has transformed diplomacy from a whispered art into a public performance—raw, rapid, and relentless.

In this world, a carefully worded policy statement competes with viral memes. A president's late-night tweet might carry more weight than a white paper. And a digital campaign can upend centuries-old institutions in a matter of weeks.

This chapter traces the transformation of social media from a public relations channel into a strategic tool—and a battleground—for diplomacy. It also explores how states have adapted to this environment, often clumsily, but sometimes with striking effect.

FROM BACKCHANNELS TO BROADCASTS

Traditional diplomacy thrived in opacity. It was a world of handshakes behind closed doors, off-the-record briefings, and deliberate ambiguity. But the digital age shattered that structure. Today, negotiations unfold in public, shaped not only by power but by perception.

Presidents announce foreign policy changes via X (formerly Twitter). Ministries livestream bilateral meetings on YouTube. Ambassadors spar with one another on Instagram. And these messages are not just seen—they are dissected, shared, parodied, and sometimes weaponized.

Broadcast diplomacy now requires diplomats to act as orators, influencers, and crisis managers—all in one scrollable package.

WHY SOCIAL MEDIA MATTERS IN DIPLOMACY

The power of social media lies not in its novelty, but in its structure. It enables a fundamentally different type of diplomatic interaction—open, decentralized, immediate, and emotionally charged.

1. Narrative Control

In today's attention economy, the first mover often defines the truth. Governments use social media to frame crises, shape narratives, and push agendas before traditional media or adversaries can respond. It's about owning the frame—not just telling the story.

2. Cultural Diplomacy at Scale

Through platforms like Instagram, TikTok, and YouTube, nations can share music, cuisine, dance, literature, and humor—tools of soft power that once took years to spread. Now, a video of traditional calligraphy or a national day celebration can garner millions of views in hours.

3. Crisis Communication

When disaster strikes—a natural catastrophe, a military attack, or civil unrest—social media becomes a lifeline. Embassies and ministries can quickly inform citizens, clarify facts, and reassure international audiences.

4. Digital Listening

Smart diplomats don't just speak. They listen. Social media provides real-time feedback on public sentiment, emerging rumors, disinformation trends, and emotional climate. Listening is now strategic intelligence.

OPPORTUNITIES: THE UPSIDE OF THE FEED

- **Speed and Accessibility:** Real-time posting enables rapid response—vital during shifting diplomatic developments.
- **Authenticity and Humanization:** Personal posts by diplomats—grief, humor, empathy—build connection and trust.
- **Direct Engagement:** Officials can challenge misinformation, respond to critics, and communicate directly with foreign populations unfiltered.

RISKS: WHEN DIPLOMACY GOES VIRAL

But with exposure comes risk. The digital world is fast, but it's also unforgiving. Missteps magnify instantly. Nuance is lost in translation.

1. Escalation by Emoji

One ill-judged comment—or worse, a tone-deaf meme—can offend, confuse, or inflame. Cultural context often vanishes in digital translation.

2. Algorithmic Incentives

Social media platforms are built to amplify outrage. Nuanced diplomacy is often buried beneath sensational posts, clickbait, or coordinated trolling.

3. Weaponized Information

Disinformation is now a geopolitical tool. Bad actors manipulate content, impersonate officials, and sow division—often without consequence.

4. Loss of Control

Once a post is live, it's no longer yours. Screenshots outlive deletions. Parodies distort intent. Messaging becomes unpredictable and difficult to reclaim.

EXPANDED CASE STUDIES: SOCIAL MEDIA AS A GEOPOLITICAL ACTOR

Case Study 1: The U.S. and Trump's Twitter Diplomacy

Introduction: A Break from Tradition

In the annals of U.S. diplomacy, President Donald J. Trump's adoption of Twitter as his megaphone marked a tectonic shift. Diplomatic messaging—once mediated through carefully composed press statements, elegant speeches, and classified cables—escaped the halls of the State Department and landed directly in the public square. Trump's Twitter account, **@realDonaldTrump**, didn't just broadcast thoughts—it shaped foreign affairs, disrupted protocol, and challenged embedded norms.

Where prior administrations leaned on teams of diplomats and advisors, Trump spoke directly—often unauthorised, late-night tweets that sometimes contradicted official policy. His Twitter feed became both a blueprint and battleground of global strategy, as financial markets, foreign capitals, and diplomats scrambled to interpret his intent.

Evolution of Presidential Communication

Early in the digital era, President Obama strategically harnessed platforms like Facebook and YouTube, but always through vetted, polished content. Trump inverted that model. His messages—posted without rehearsal—felt like personal broadcasts, initiating policy shifts through unfiltered dialogue. He turned tweets into directives, pronouncements, or provocations, redefining how presidential messaging worked.

This fundamentally shifted the diplomatic landscape. Foreign embassies retooled: every incoming message from Trump's account became 'must-read intelligence'. Seminars on social media analysis replaced sessions on cable drafting. Diplomacy was now live.

Twitter Diplomacy in Action: Key Examples

1. North Korea — Bluster to Summit

In 2017, tensions with Pyongyang reached fever pitch. Trump unleashed sharp-tongued tweets—lambasting Kim Jong-un as "Little Rocket Man" and threatening "fire and fury." The global community panicked, expectations darkening. But unexpectedly, this posture led to thaw: summits in Singapore (2018) and Hanoi

(2019). Critics labeled his style erratic; defenders saw performance as a strategic pivot. Either way, diplomacy took place in public view and on Twitter.

2. Iran — Sanctions at 2 a.m.

In a dramatic tweet, Trump warned Iranian President Rouhani in all caps: *"NEVER, EVER THREATEN THE UNITED STATES AGAIN."* Bond markets reacted, oil prices wavered, and the rial cratered—all before dawn in Washington. His posts weren't just rhetoric—they were financial and geopolitical cues, often delivered without warning.

3. China—Tariffs from the Tweets

The trade war with China unfolded in public real-time, tweet by tweet. Trump's tariff announcements—suddenly and without briefing—triggered global market tremors and diplomatic confusion. Beijing began parsing tweets as unilateral policy, not private bluster. For business, stability shifted to social media.

Fallout: Allies, Adversaries, Upended Norms

The tweet-as-policy tagline forced governments to adapt. Allies were blindsided; ambassadors watched feeds instead of attending briefings. One Pentagon official confessed: "We learned of troop decisions from Twitter." Institutions were left to backfill reality after the tweets.

Trump's approach also reshaped global norms. Influential leaders—including Modi, Netanyahu, and Bolsonaro—shifted to public-facing platforms, conscious that diplomacy had gone viral. Subtlety gave way to soundbite geopolitics.

Analysis: Power, Pitfalls, Precedents

- **Power Amplified:** Trump showed how direct digital engagement could cut through media filters and steer conversations.
- **Strategic Blindside:** Impulsive tweets disrupted coordination—alliances weakened, policies scrambled, allies anxious.
- **Norms Rewritten:** Diplomacy became performative and public. Longstanding norms of discretion and careful messaging unraveled.

Lessons for Future Diplomacy

- **Strategic coherence beats spontaneity.** Digital messaging must align with policy infrastructure.

- **Authenticity is a double-edged sword.** A human tone engages—but risks alienation and error.
- **Institutions must evolve.** Monitoring, interpreting, and responding to digital outreach requires modern capacity and protocol.

Conclusion:

Legacy of a Tweet-Powered Presidency

Trump's Twitter diplomacy shattered conventions and triggered a global experiment in digital statecraft. Whether regarded as a coup or chaos, it demonstrated that tweets carry weight—they can unsettle markets, initiate talks, and reshape alliances. As future leaders speak live and loud, diplomacy must be ready—not reactive. Why? Because now, every presidential tweet is a policy act.

Case Study 2: The Arab Spring and the Power of Facebook

How a Platform Became a Political Force

Introduction:

The Revolution Will Be Livestreamed

In late 2010 and early 2011, a wave of civil uprisings swept across the Middle East and North Africa. Known collectively as the **Arab Spring**, these protests began without a unified ideology, political party, or common leader. What they shared was frustration—at authoritarianism, corruption, unemployment, and economic stagnation. But more than anything, they shared a **digital lifeline: Facebook**.

Though many still debate the true role of social media in these revolutions, one fact is clear: **Facebook became a central nervous system for dissent**. In a region where press freedom was tightly controlled and traditional opposition silenced, social media enabled citizens to mobilize, document, and amplify their voices in real time.

Facebook transformed from a place of friendship and photos to a platform of resistance, coordination, and citizen journalism. It wasn't built for revolution—but it became a revolutionary tool.

Tunisia: A Spark Goes Viral

It started in the small town of Sidi Bouzid, Tunisia. On December 17, 2010, Mohamed Bouazizi, a street vendor, set himself on fire after being harassed and humiliated by municipal officials. His act of despair—deeply symbolic in a nation suffocating under inequality and authoritarian rule—could have gone unnoticed. But a bystander recorded the event on a mobile phone and posted it to **Facebook**.

The video went viral across Tunisia, bypassing state censors and igniting protests in city after city. Images of demonstrators, police violence, and raw emotion flooded timelines. Within a month, **President Zine El Abidine Ben Ali**—who had ruled for 23 years—fled the country.

The Tunisian revolution was the first of its kind: a leader deposed, in part, by a **digitally coordinated uprising**.

Egypt: When a Page Becomes a Movement

The revolution in Egypt remains one of the most iconic examples of Facebook's power in political mobilization. The pivotal moment came with the creation of a single Facebook page: **"We Are All Khaled Said."**

Khaled Said, a 28-year-old man from Alexandria, was beaten to death by police in 2010. Graphic photos of his disfigured face circulated online. In response, digital activist **Wael Ghonim**, then a Google employee, created the page anonymously. It soon became a symbol of Egypt's police brutality and impunity.

On January 25, 2011—National Police Day—calls went out via Facebook to protest. The platform's reach was so effective that tens of thousands gathered in **Tahrir Square** in Cairo, sparking weeks of massive protests, which eventually led to the resignation of **President Hosni Mubarak** on February 11 after 30 years in power.

When the government tried to shut down the internet entirely, activists circumvented the blackout with **proxy servers, international SIM cards**, and the help of the global diaspora. Digital resilience became just as vital as physical resistance.

Libya, Syria, Yemen: From Protest to War

In nations with weaker state institutions and more entrenched authoritarian regimes, Facebook and other platforms played a different—but equally powerful—role.

- **Libya:** Protest videos uploaded by civilians exposed Muammar Gaddafi's brutal crackdowns. The international community, prompted in part by digital footage, responded with a NATO military intervention.
- **Syria:** What began as peaceful protests in Daraa escalated into civil war. Videos of regime violence posted to Facebook and YouTube shocked the world but also marked the start of a catastrophic conflict. Citizen journalists became targets.
- **Yemen:** Activists used Facebook to organize hunger strikes, youth protests, and sit-ins. Social media played a major role in toppling President Ali Abdullah Saleh, although the post-revolution era quickly unraveled into civil war.

Across these countries, Facebook acted as both a mirror and a megaphone—reflecting oppression and amplifying resistance. But it also exposed activists to surveillance and retaliation.

The Double-Edged Sword of Digital Uprising

A. Empowerment through Connectivity

Social media allowed revolutions to become leaderless yet coordinated. Posts, hashtags, and livestreams enabled real-time decisions and crowd-sourced strategies. An event in Cairo could inspire marches in Damascus or spark solidarity in Manama. Images became fuel for action, and borders became irrelevant.

B. Platforms as Political Players

While Facebook claimed neutrality, its very infrastructure shaped the revolutions. By enabling visibility and connection, it became a stakeholder in political outcomes. However, the platform was never designed for governance, truth moderation, or conflict mediation. Its algorithms favored virality over verification—fueling polarization, misinformation, and emotional extremism.

C. From Liberation to Surveillance

Authoritarian regimes adapted quickly. They began using Facebook to track dissidents, infiltrate groups, and spread disinformation. Activists' digital fingerprints made them vulnerable. In several cases, footage shared online was used in trials against protesters or in targeting operations.

D. Western Idealism Meets Regional Complexity

Western observers initially hailed the Arab Spring as a digital democratization wave. But enthusiasm soon met reality. Institutional vacuum, sectarian tensions, and foreign interference derailed many transitions. Egypt returned to authoritarian rule. Syria descended into prolonged war. Libya fragmented. The digital victory proved far easier than democratic consolidation.

Lessons Learned

- **Social media can start a revolution, but not finish it.** Platforms are excellent for coordination and mobilization—but poor at governance and institution-building.

- **Visibility is power—and risk.** What makes social media effective also makes it dangerous in authoritarian settings.
- **Tech platforms are now geopolitical players.** Whether or not they want the responsibility, they must develop policies, infrastructure, and accountability mechanisms to manage their influence on political life.

Conclusion: The Algorithmic Awakening

The Arab Spring proved that **Facebook could topple governments**—but it also showed that **technology alone cannot deliver democracy**. While the revolutions brought fleeting moments of unity, they also revealed the limits of networked activism in the absence of structural change.

Still, the world had changed. The hashtag had joined the ballot box. The livestream had joined the protest. And the digital public square had become an arena of power.

The legacy of the Arab Spring lives on—not just in the region's political scars, but in the global understanding that **no regime, no platform, and no system is immune to the viral force of public will.**

Case Study 3: Ukraine's Digital War Room

Fighting on the Frontlines of Information

Introduction:

The First War of the Information Age

When Russia launched its full-scale invasion of Ukraine in February 2022, the world braced for classic warfare: tanks, missiles, and human suffering. Yet alongside the kinetic battlefield, a different but equally decisive conflict emerged—one fought with smartphones, social media, and digital strategy. President Volodymyr Zelenskyy and his team pioneered a form of modern resistance—embedding digital communication into military and diplomatic tactics—creating what came to be known as **Ukraine's digital war room.**

In this new hybrid war, every tweet, meme, and video was weapons-grade. Swift messaging shaped international perceptions, mobilized resources, and countered enemy narratives. The conflict raised an urgent question: can a nation survive—and thrive—on digital frontlines? Ukraine answered with resounding urgency: yes.

Leadership in the Lens

On February 24, 2022, Kyiv awoke to sirens. President Zelenskyy's response was immediate—and intimate. In a low-resolution video filmed inside Kyiv, he stood alongside aides and calmly spoke: *"We are all here. Our country is fighting. We will win."*

His tone—steady, emotional, and unfiltered—stood in stark contrast to scripted presidential addresses. His bravery was punctuated not by force, but by proximity: Zelenskyy's camera shook like the city, not like a teleprompter.

This approach reframed international support. Audiences didn't just hear Ukraine's story—they felt it. Parliaments and NGOs responded. Solidarity campaigns and grassroots fundraising blossomed online. Zelenskyy's leadership style—digitously human—emerged as both a psychological tool and a diplomatic signal.

Meme Warfare: Humor as Resistance

Amidst the horror, humor became a weapon of defiance. Ukraine's public communications team, bolstered by citizen volunteers, launched rapid-fire memes dismantling Russian propaganda and highlighting the absurdity of aggression. One of the most viral lines **"Russian warship, go f**k yourself"** captured global attention, becoming a symbol of both defiance and Ukrainian ingenuity.

Memes served multiple purposes:

- **Boosting morale** for soldiers and citizens alike.
- **Internationalizing the struggle** through shareable content.
- **Undermining enemy narratives**, turning fear into ridicule.

The result was contagious resilience. The world saw not just another conflict—but a spirited struggle between justice and injustice, between David and Goliath.

Mobilizing Digital Citizens

Ukraine's digital mobilization went beyond storytelling—it activated digital citizens:

1. Crowdfunding for defense and aid
Platforms like GoFundMe, Kickstarter, and Patreon powered grassroots defense funds. From tactical

drones to school supplies, online donors supplied Ukraine with vital equipment—directly bypassing bureaucratic slowdowns.

2. Cyber defense and volunteer hacking

As Russian cyber-attacks targeted critical infrastructure, Ukraine's tech volunteers rallied. National and international coders formed IT armies, patching systems, shielding utilities, and launching counter-cyber strikes. These decentralized actors blurred the line between civilian activism and national defense.

3. Behind-the-scenes tech diplomacy

Zelenskyy's team targeted tech giants directly. Microsoft, Google, Apple, and Netflix were called on to suspend operations in Russia. In just days, multiple platforms pulled services—demonstrating how digital shaming could translate into corporate pressure.

Sustaining Global Solidarity Through Storytelling

Zelenskyy adapted storytelling for every audience. He addressed the U.S. Congress, Canada's Parliament, and European assemblies via livestream—not from an embassy, but from Kyiv. His words were amplified by visuals: children in bomb shelters, humanitarian workers in trenches. He turned micro-narratives into macro-movements.

This bolstered international resolve:

- **Sanctions hardened**, with unified Western blocs banning Russian elites and banks.
- **Military aid surged**, as public pressure influenced policy.
- **Global public opinion** remained empathetic, sustained by daily digital updates.

Ukraine reframed itself—not as victim, but as guardian of democratic values.

Digital Warfare: Lessons for Tomorrow

1. Authenticity commands attention

Authentic, low-production messaging can carry higher emotional impact than polished broadcast.

2. Strategic narrative is national defense

Ukraine's story-building shaped battlefield conditions and diplomatic relations.

3. Citizen-led action matters

Cyber volunteers and crowdfunders became integral participants in national strategy.

4. Digital resilience is geopolitical

Cloud infrastructure, connectivity networks, and secure communication became as vital as frontline trenches.

Conclusion: The Digital Frontline Endures

Ukraine's digital war strategy redefined modern warfare. In a conflict where geography and firepower once decided outcomes, now **connectivity and communication** can determine survival. Zelenskyy's team demonstrated that digital tools aren't merely supportive—they're strategic. They built a new taxonomy of defense—where open-source intelligence, viral resistance, and collective mobilization converge into unified action.

As global threats evolve—from cyber espionage to disinformation campaigns—Ukraine stands as a living case study: when your adversary bombs your cities, respond with bandwidth, bravery, and believable storytelling.

The digital war room has arrived. And it is as consequential as any battlefield sortie.

CONCLUSION: DIPLOMACY BY FEED, NOT FAX

We've entered an era where diplomacy no longer unfolds solely in closed-door negotiations or through carefully crafted communiqués—it now lives, breathes, and breaks in the feed.

Platforms like X (formerly Twitter), TikTok, Telegram, and even Reddit are no longer on the periphery of geopolitics. They are central stages where influence is asserted, legitimacy is tested, and global narratives are contested in real time. Social media has become the new diplomatic theater—public, unfiltered, and immediate.

This shift has not replaced traditional diplomacy but rather layered a volatile and high-velocity dimension on top of it. What was once a calculated process of weeks can now collapse—or crystallize—in moments. Leaders are expected to react not by statement, but by post. A misjudged emoji, a delayed response, or an overly aggressive thread can disrupt alliances, inflame tensions, or inadvertently signal intent.

The diplomat of the digital age must master far more than treaties and protocol. Today's diplomatic toolkit includes narrative framing, crisis messaging, algorithmic literacy, and real-time ethical decision-making. The stakes are enormous. A mistimed tweet can unravel years of trust. A well-placed post can rally nations or humanize a crisis.

Yet with this power comes vulnerability. The same tools that amplify truth can spread disinformation. The same platforms that connect can also polarize. Diplomacy, once guarded by distance and delay, now operates in the glare of virality—where judgment is swift and memory is forever.

As we turn the page, we leave the realm of influence and messaging and enter a darker, more fragile frontier: the infrastructure beneath it all. In the next chapter, we examine how diplomacy is now being forced to defend not just its narratives—but its very digital foundations—in the escalating domain of cybersecurity.

Chapter 4: War Without Borders

CYBERSECURITY AND THE BLURRED LINES BETWEEN ESPIONAGE, WAR, AND DIPLOMACY IN THE VIRTUAL DOMAIN

In the modern age, war doesn't always begin with boots on the ground or bombs from the sky. Sometimes, it starts with a glitch. A power outage. A leaked database. Or the silent collapse of a nation's digital backbone. Welcome to the era where diplomacy no longer protects just people and borders—but networks, code, and data.

Cybersecurity is no longer a niche concern reserved for IT professionals. It is now a central pillar of foreign policy and a defining frontier of global diplomacy.

In this chapter, we explore how diplomacy is evolving in response to the growing wave of cyber threats—and how nations are forging new alliances, norms, and strategies to protect their sovereignty in a world that's always connected, and always vulnerable.

THE RISING IMPORTANCE OF CYBERSECURITY IN INTERNATIONAL RELATIONS

Twenty years ago, cyberattacks were mostly confined to the realm of espionage and criminal hacking. Today, they're strategic instruments of statecraft. Nation-states use cyber capabilities to steal secrets, sabotage rivals, disrupt infrastructure, and shape public perception.

And unlike traditional warfare, cyber warfare is deniable, borderless, and incredibly difficult to attribute. A malware infection halfway around the world might be the work of a government—or a lone actor on a laptop in a basement. Either way, the damage is real.

From election interference to ransomware attacks on hospitals, from energy grid takedowns to fake news campaigns, cyberspace is no longer just a tool—it is a contested domain. And diplomats are now expected to navigate it.

DEFINING CYBER DIPLOMACY

Cyber diplomacy is the practice of using diplomatic channels and frameworks to manage international cybersecurity issues. It includes:

- Negotiating norms of behavior in cyberspace.
- Creating trust-building mechanisms among states.
- Managing cyber incidents to prevent escalation.
- Securing digital infrastructure across borders.
- Advocating for rights-based digital governance.

It's about applying the logic of diplomacy—dialogue, deterrence, trust, and multilateralism—to the digital battlefield.

KEY ELEMENTS OF CYBER DIPLOMACY

1. International Norm-Setting

Because cyberspace transcends borders, a patchwork of rules creates dangerous ambiguity. That's why cyber diplomacy prioritizes global norms—rules of the road to determine what constitutes acceptable state behavior.

- Should a country be allowed to hack into another nation's power grid during peacetime?
- Is election interference via social media a violation of sovereignty?
- What responsibility do states have to control cybercriminals operating within their borders?

Multilateral efforts—through the UN, regional alliances, and digital cooperation forums—are slowly shaping answers. But consensus remains difficult.

2. Attribution and Accountability

One of the thorniest issues in cyber diplomacy is determining who is responsible for an attack. Unlike a missile strike, a cyberattack often leaves unclear fingerprints. Even if a state is suspected, proving intent is murky.

Without clear attribution, retaliation risks becoming miscalculated—and escalation more likely.

Cyber diplomats must balance technical evidence with strategic patience. Often, they must choose between public condemnation and backchannel deterrence.

3. Confidence-Building Measures (CBMs)

Just as Cold War diplomats developed communication hotlines to avoid nuclear missteps, cyber diplomacy increasingly relies on CBMs—agreements between countries to share information, notify each other of major incidents, or avoid targeting critical civilian infrastructure.

These trust-building steps don't prevent every attack, but they lower the chances of misinterpretation.

4. Capacity Building and Digital Solidarity

Cyber resilience is not evenly distributed. Smaller or developing nations often lack the expertise and resources to protect their digital systems.

Cyber diplomacy includes development assistance—helping countries build secure digital infrastructure, training cybersecurity professionals, and creating regional cybersecurity centers.

This isn't just altruism—it's strategic. Weak links in global cybersecurity can become backdoors for global threats.

CASE STUDY 1: ESTONIA – THE DIGITAL REPUBLIC UNDER SIEGE (2007)

"The First Cyber War" and the Emergence of a Digital Defense Doctrine

In the spring of 2007, the streets of Tallinn were quiet. But behind the scenes, a different kind of chaos was unfolding—one not visible to the naked eye. For a country known for its pioneering digital infrastructure, Estonia suddenly found itself in the middle of something unprecedented: a full-scale cyberattack. There were no tanks, no missiles, no military uniforms. The battleground was digital, the weapons were lines of code, and the consequences were very real.

Estonia affectionately nicknamed "E-stonia," had built a global reputation as a digital trailblazer. Citizens voted online. Prescriptions were filled electronically. Taxes were filed in minutes. The entire country had embraced e-governance not just as a convenience, but as a core expression of national identity. Innovation was Estonia's proud badge of honor—and its Achilles' heel.

It all began with a decision rooted in national memory and unresolved trauma. The Estonian government announced it would relocate the Bronze Soldier—a Soviet-era statue commemorating Red Army soldiers—from central Tallinn to a military cemetery. To many ethnic Estonians, it was a long-overdue symbolic act of reclaiming their historical narrative from the Soviet past. But to the Russian-speaking minority, and to Moscow itself, it was an insult—an erasure of anti-fascist heroism.

Protests erupted. Tensions flared. And then, the invisible war began.

What started with outrage in the streets quickly escalated into an all-out digital siege. In late April 2007, Estonia's government websites began crashing. At first, it seemed like a glitch. Then the banks followed. Then media outlets. Then the emergency services. Within days, it was clear: this was a coordinated cyber assault.

The Anatomy of a Digital Attack
The attacks were relentless. Distributed Denial of Service (DDoS) campaigns flooded servers with junk traffic, overwhelming critical infrastructure.

- Government websites, including those of the prime minister and parliament, were rendered useless.
- Media outlets found their digital platforms frozen—especially those that criticized Russian policies.
- Banks couldn't process transactions, and ATMs went offline across the capital.
- Even basic email communications between state agencies were disabled.

The cyber onslaught lasted for weeks. Much of the traffic came from IP addresses scattered across the globe, but many were traced back to Russia. Though Moscow denied any official involvement, the scale and coordination pointed to something more than just patriotic hackers. It felt like a test—a warning shot in a new kind of geopolitical theater.

Between Bullets and Bytes
The attack left Estonia paralyzed. But it also left the international community in a legal and diplomatic vacuum. There were no protocols for this. No doctrine. No playbook.

Under NATO's founding treaty, Article 5 states that an attack on one member is an attack on all. But what constitutes an "attack" in cyberspace? Estonia appealed to its allies, but even among friends, confusion reigned. Was this war? Sabotage? Or something in between?

Estonian officials began lobbying for a broader understanding of collective defense—one that included digital warfare. At the same time, they scrambled to secure their systems, calm the public, and launch a global conversation that would eventually reshape NATO's posture on cybersecurity.

Turning Crisis into Capacity

Estonia didn't just recover—it adapted. With remarkable speed and clarity, it transformed itself from a digital victim into a cyber diplomacy pioneer.

- The government launched widespread public education campaigns on digital hygiene and threat awareness.
- Significant investments were made in segmented networks, redundant systems, and rapid-response protocols.

- Estonia hosted international cyber defense forums, calling for legal frameworks to guide state behavior in cyberspace.
- It advocated for the establishment of dedicated cybersecurity centers within NATO and the EU.

In 2008, just a year after the attacks, the NATO Cooperative Cyber Defence Centre of Excellence (CCDCOE) was established in Tallinn. It became a hub for cyber policy, military simulations, and multinational collaboration—cementing Estonia's place at the forefront of global cyber defense.

Why This Case Still Matters

The Estonia case was a turning point. It forced the world to confront uncomfortable questions:

- Can a cyberattack be considered an act of war?
- What does sovereignty mean when your borders are breached not by troops, but by data packets?
- How should alliances respond when deterrence isn't defined by missiles, but by malware?

Estonia, a nation of just over a million people, taught the world that size doesn't matter in digital warfare—

resilience does. It showed that a small, agile state with the right tools and strategic vision could influence global norms.

Diplomacy in the Age of Firewalls

Perhaps most importantly, Estonia helped give birth to cyber diplomacy—a field where embassies are replaced by encryption, and international law must stretch to meet new threats. It advocated not just for cybersecurity, but for cyber accountability. And it urged the international community to recognize that the next great conflict might not begin on a battlefield—but on a server farm.

CASE STUDY 2: U.S. – CHINA CYBER TENSIONS AND THE 2015 XI-OBAMA AGREEMENT

From Digital Espionage to Diplomatic Détente—
Briefly

The war wasn't loud. It didn't play out on CNN or through tanks rolling across borders. It was quiet, surgical, and invisible to most of the world. But in the corridors of Washington and the command centers of

Beijing, it was war nonetheless—waged not with bullets, but with code.

By the early 2010s, a storm had been brewing between the United States and China—not on trade tariffs or military exercises, but in cyberspace. U.S. officials were sounding alarms about massive cyber espionage campaigns they believed were being carried out by Chinese state-backed actors. These weren't speculative attacks—they were calculated, large-scale intrusions targeting everything from top defense contractors to pharmaceutical firms and government databases.

And then, in 2015, the storm broke. A single breach turned quiet frustration into public confrontation. That breach was the Office of Personnel Management (OPM) hack, and it marked a watershed in cyber diplomacy.

Prelude to Crisis: A Pattern of Digital Intrusions
This wasn't an isolated incident. The OPM breach came after years of mounting cyber aggression attributed to Chinese hackers.

- In 2009, Operation Aurora made headlines when Google and dozens of other U.S. tech companies were infiltrated, their source code compromised.
- Cybersecurity firms like Mandiant (later acquired by FireEye) published reports detailing Chinese military-linked cyber units like APT1, which allegedly operated out of a nondescript building in Shanghai.
- Industries from aerospace to clean energy to pharmaceuticals were being regularly raided—digitally—for their intellectual property.

Despite repeated diplomatic protests, Washington felt stuck. These were not acts of war, technically. But the theft was costing billions and undermining national security. For years, the conflict sat in a geopolitical gray zone: too shadowy for open retaliation, too serious to ignore.

The Spark: The OPM Breach and Its Implications
Then came the breach that changed everything.

In June 2015, it was revealed that the U.S. Office of Personnel Management—a government agency responsible for handling security clearances—had been hacked. And not just slightly hacked. Over 21 million personal records had been compromised.

The data included:

- Fingerprints
- Financial histories
- Family and social connections
- Psychological evaluations

This was not a simple identity theft operation. This was a national security catastrophe. Intelligence officers, military personnel, diplomats—many now found their entire personal and professional lives potentially exposed to a foreign power. In the intelligence community, it was feared that China could use this information to map networks, track agents, and exploit vulnerabilities for years to come.

In Washington, pressure mounted. Some called for sanctions. Others warned against escalation. But one thing was clear: the old "ignore it and move on" strategy could no longer hold.

Diplomacy on the Edge: The Xi–Obama Summit

Just three months after the OPM breach, Chinese President Xi Jinping arrived in Washington for a state visit. All eyes were on the summit. Many wondered: would the two countries trade barbs—or strike a deal?

Surprisingly, they chose the latter.

In a move that stunned observers, Presidents Obama and Xi stepped in front of the cameras and announced a bilateral agreement on cyber conduct—a first between two global rivals.

The agreement included:

- A **mutual pledge** not to conduct or support cyber-enabled theft of intellectual property for commercial advantage.
- The creation of a **Cyber Working Group** to improve transparency and build trust.
- Plans for **law enforcement cooperation** between the FBI and China's Ministry of Public Security.
- A commitment to **ongoing dialogue**, including confidence-building measures to prevent miscommunication and escalation.

For many, it was the first real acknowledgment that cyber diplomacy had arrived. Cyberspace, long treated as a technical or military concern, was now officially part of global statecraft.

The Short-Lived Success of Cyber Détente
In the months that followed, something unusual happened. U.S. cybersecurity firms began reporting **a decline in Chinese-origin cyber intrusions**, particularly those targeting commercial IP. While some remained skeptical—was it simply better obfuscation?—the evidence suggested Beijing was, at least temporarily, scaling back.

There were also new signals of cooperation:

- China began prosecuting some of its own cybercriminals.
- Joint law enforcement meetings took place.
- Dialogue channels opened between the two countries' cybersecurity officials.

For a moment, it seemed like diplomacy might work in cyberspace.

The Unraveling: New Politics, New Tensions
By 2017, with the Trump administration in office, the tone toward China had shifted dramatically. The U.S. adopted a more combative stance, citing national security threats linked to Chinese technology firms like Huawei. Trade wars escalated. Accusations flew. And in cyberspace, the gloves came off.

The cyber working group quietly dissolved. New cyberattacks resumed, often more sophisticated and less traceable. The sense of cautious optimism from 2015 was gone. The Xi–Obama cyber agreement had faded—not with a bang, but with a shrug.

Why This Moment Still Matters
Despite its brief lifespan, the 2015 U.S.–China cyber accord marked a turning point. It proved that even bitter rivals could come to the table and define behavioral norms in the digital realm. It showed that cyberspace was not immune to diplomacy—it required it.

More importantly, it reframed cyberattacks as more than just "IT issues." They became **national security, economic, and diplomatic concerns**—worthy of summit-level negotiation.

Lessons from the Brink

- **Attribution is critical.** For cyber diplomacy to work, states need reliable forensics and transparency.
- **Cyber diplomacy requires structure.** Ad hoc deals, no matter how historic, need enforcement mechanisms to survive leadership changes.
- **Digital peace is possible—but temporary.** In an era of techno-nationalism, trust must be constantly negotiated.
- **Diplomatic tools must evolve.** Codes of conduct, incident hotlines, and cyber norms should be institutionalized—just like arms control frameworks.

Conclusion: The Beginning of a Conversation

The Xi–Obama agreement may not have endured, but it changed the conversation. It elevated cyberspace into the realm of diplomacy—not as a fringe concern, but as a central pillar of international security.

In hindsight, the agreement was less about ending conflict and more about managing rivalry. It showed that in the digital age, diplomacy doesn't end when the hack is discovered—it begins there.

CASE STUDY 3: AFRICAN UNION HQ – THE GREAT DATA HEIST (2012-2017)

A Global Lesson on Digital Trust and Sovereignty

In 2012, the African Union unveiled a modern architectural marvel in the heart of Addis Ababa—a soaring glass-and-concrete headquarters that gleamed with ambition. The $200 million building, gifted by the Chinese government, stood not just as a symbol of pan-African unity, but as a testament to a new era of Sino-African cooperation. Designed, constructed, and outfitted by Chinese firms, it was hailed as a gesture of generosity from a rising global power toward a continent seeking infrastructure, investment, and global standing.

But what looked like diplomacy in stone and steel would, five years later, become the centerpiece of one of the most unsettling digital breaches in modern diplomatic history.

The Revelation: A Midnight Stream to Shanghai

In January 2018, French investigative journalists from Le Monde Afrique broke a stunning story. For nearly five years—between 2012 and 2017—sensitive data from the AU's headquarters had been siphoned off nightly. Every evening between midnight and 2 a.m., servers inside the AU complex silently uploaded massive volumes of internal communications and documents to unknown destinations. Eventually, investigators traced the data flows—not to an African capital or third-party proxy—but to servers in **Shanghai, China**.

According to the report:

- The breach had gone undetected for years.
- The digital infrastructure—servers, software, and systems—was fully installed and managed by Chinese contractors.
- Logs and audits revealed consistent unauthorized data extraction during off-hours.

This wasn't a rogue intrusion. It was quiet, systemic, and devastating in its implications.

The Silence That Followed

The African Union's response was swift—but muted. There were no dramatic press briefings or formal accusations. Officials spoke in generalities. China's foreign ministry dismissed the report as "absurd" and accused critics of undermining Sino-African friendship.

Behind closed doors, however, the mood was different. Senior African diplomats and IT officers scrambled to replace servers, reinforce network firewalls, and shift key communication protocols. Cybersecurity specialists were flown in to conduct silent audits. Contracts with Chinese tech providers were reconsidered—but without fanfare or public fallout.

There were no sanctions. No diplomatic expulsions. Just a deep recalibration cloaked in discretion.

Why the Silence? A Diplomatic Balancing Act

The breach presented African leaders with a difficult choice: defend digital sovereignty or preserve critical economic relationships?

China, after all, was not merely a donor. It had become the continent's **largest trading partner**, investing billions in infrastructure projects—from railways and roads to energy grids and smart cities. To accuse China publicly would risk derailing deals, projects, and diplomatic goodwill built over decades.

Instead, African governments chose a different path: quiet resilience over noisy confrontation. What followed was a continental awakening:

- The AU commissioned a **complete replacement of its IT systems** using non-Chinese suppliers.
- Several countries initiated **independent audits** of Chinese-built data centers, smart city technologies, and surveillance systems.
- Discussions around **data localization** and African-led cloud services gained momentum.

The scandal, though never formally named as such, reshaped Africa's approach to digital sovereignty from the inside out.

Infrastructure Is Not Neutral
This case exposed an uncomfortable truth: who builds your infrastructure often retains invisible levers of access. The cables under your city, the software in your servers, the companies that maintain your networks—

they can all serve double functions: utility and influence.

In the case of the AU, the very heart of continental diplomacy had become a **strategic listening post**, constructed by a state actor whose geopolitical ambitions spanned far beyond aid and infrastructure.

What emerged was a new awareness:

- **Digital infrastructure is strategic infrastructure.** It cannot be outsourced without due diligence.
- **Cybersecurity must be designed from day one**—not retrofitted in response to scandal.
- **Procurement processes matter.** Contracts for high-level IT systems must include transparency, local oversight, and clauses around data sovereignty.
- **Silence can be strategic.** Sometimes, discretion buys time to recalibrate, protect reputations, and rebuild trust.

Contextualizing the Breach: China's Digital Silk Road. The African Union case didn't happen in a vacuum. It was part of a broader trend tied to China's **Digital Silk Road**—a global initiative to export

Chinese-built digital infrastructure across Asia, Africa, and Latin America.

Through companies like Huawei and ZTE, China was not just selling hardware—it was shaping standards, embedding software, and offering bundled "smart governance" solutions to governments around the world.

Africa, with its vast need for digital connectivity and limited local capacity, became an ideal testing ground. But as the AU breach demonstrated, dependency can come at a cost—not always financial, but strategic.

Diplomacy Without a Microphone
What makes this case so distinct is what didn't happen. There was no public showdown. No headlines about diplomatic expulsions or emergency summits. Instead, the response took place behind closed doors—in server rooms, procurement offices, and cybersecurity task forces.

This was diplomacy conducted in silence. But it was diplomacy, nonetheless.

It showed that in the digital age:

- Retaliation is not always loud.
- Trust, once lost, is recalibrated in policies, not press releases.
- Sovereignty is increasingly exercised through **technology choices**, not territorial declarations.

A Lesson in Modern Sovereignty

The African Union data breach became a turning point—not just in cybersecurity, but in how digital trust is managed between states.

Africa emerged from the experience more cautious, more assertive, and more committed to shaping its own digital destiny. From revised procurement rules to investments in homegrown cybersecurity expertise, the continent began laying the groundwork for more autonomous digital infrastructure.

Conclusion: A Gift with Hidden Terms

What began as a symbol of unity between Africa and China ended as a cautionary tale. The AU building was not just a diplomatic headquarters—it became a lesson in 21st-century power, where access to information is as valuable as territory, and trust can be breached not with soldiers, but with silent scripts running at midnight.

Today, the AU still stands tall in Addis Ababa. But its digital systems have changed, its contracts are reviewed more carefully, and its leaders are more aware than ever that **sovereignty is wired—not just written.**

This case reminds us: in a world where connectivity is currency, the ultimate question is no longer "Who funds your infrastructure?"—but **"Who controls it?"**

CONCLUSION: CYBER DIPLOMACY AT THE EDGE OF STABILITY

Diplomacy today unfolds in a world where borders are invisible, adversaries are anonymous, and digital risks are escalating with unprecedented speed. In this shifting landscape, **cyber diplomacy is no longer optional—it is essential.**

The case studies explored in this chapter—from **Estonia's digital awakening** and the **fragile U.S.–China cyber accord**, to **Africa's quiet recalibration**—all converge on a sobering reality: **the**

next war may be fought with missiles, but it will begin with malware.

Cyber threats have fundamentally redrawn the map of diplomacy. Unlike conventional warfare, digital attacks are often **unattributed, unannounced, and legally undefined**. They strike not on battlefields, but in the heart of civilian life—crippling hospitals, disrupting water systems, stealing personal data, and eroding trust in public institutions.

And yet, diplomacy has not disappeared—it has evolved. It has become quieter, more agile, and more layered. Backchannels, expert working groups, emergency hotlines, and silent crisis rooms have emerged to manage escalation, negotiate norms, and avert catastrophe. In many cases, the most successful act of diplomacy is not what is said—but what is prevented: the attack that doesn't escalate, the retaliation that doesn't materialize, the breach that prompts coordination, not collapse.

The demands placed on 21st-century diplomats are immense:

- To **negotiate norms in a domain without geography**.
- To **build trust where truth is elusive and attribution uncertain**.
- To **collaborate globally while competing strategically**.
- And to **defend civilians from threats that bypass traditional diplomatic frameworks—but strike at the core of sovereignty**.

To meet these demands, cyber diplomacy must now rest on four foundational pillars:

- **Clarity** – in defining what constitutes aggression, violation, and legitimate defense in cyberspace.
- **Coordination** – between governments, private sector actors, multilateral bodies, and civil society.
- **Capacity** – to ensure all states, regardless of size, can defend their digital sovereignty and contribute meaningfully to cyber governance.
- **Courage** – to lead, to set norms, and to act responsibly in a space where few rules yet exist.

The stakes could not be higher. As digital infrastructure becomes as critical to national interest as ports, embassies, or airspace, the fragility of our interconnected systems becomes starkly visible. A single misconfigured server, a rogue keystroke, or a misinterpreted scan could ignite a chain reaction with consequences far beyond the digital domain.

But this moment also carries opportunity. Cyber diplomacy can become a great equalizer, offering smaller nations new agency in global affairs. It can foster transparency, coordination, and resilience. And if approached with humility and vision, it can even help rebuild trust between adversaries—one line of code, one quiet agreement at a time.

We are still in the opening chapters of this new era. The doctrine is being developed in real time. But one truth is already clear: **cyber diplomacy is no longer a technical specialization—it is frontline statecraft. It belongs not to a department, but to the mission of every nation.**

As we move forward, the next chapter explores the **preparation required for this reality:** how the diplomats of tomorrow must be trained, equipped, and empowered for a world where power is digital, conflict is asymmetric, and diplomacy must be as agile as the networks it seeks to secure.

Chapter 5: Information Warfare: Shaping Narratives Digitally

THE MOST POWERFUL WEAPON OF THE 21ST CENTURY DOESN'T EXPLODE; IT PERSUADES

In today's world, power is no longer measured only in troops, tanks, or treaties. It is measured in influence—who controls the story, who shapes the headlines, who captures the scroll of the screen. Wars are still fought on land, sea, and air—but increasingly, the battlefield is the mind.

This is the age of information warfare—a struggle not for physical domination, but for narrative supremacy. It is a war of images, memes, manipulated truths, selective facts, and carefully curated doubt.

And the frontlines are everywhere: social media feeds, encrypted chat groups, state-run news outlets, troll farms, livestreams, and algorithm-driven search results.

This chapter explores how information is being weaponized—and what it means for diplomacy, democracy, and global stability.

THE RISE OF THE NARRATIVE BATTLEFIELD

In previous generations, propaganda was centralized. Leaflets, radio broadcasts, tightly controlled media. Today, it's decentralized, personalized, and frictionless.

- Disinformation spreads:
- Faster than traditional news
- Through trusted personal networks
- With emotional precision

What was once a state function is now an **ecosystem**, where nation-states, private actors, bots, influencers, and ordinary citizens can all become **amplifiers of truth—or deception**.

THE NEW TOOLS OF INFORMATION WARFARE

1. Memes and Micro-Messaging

A well-timed meme can do more than a thousand-word policy paper. Humorous, relatable, and emotionally charged, memes have become cultural weapons used to erode credibility or build solidarity.

2. Deepfakes and Synthetic Media

AI-generated videos and voice clips can now fabricate speeches, confessions, or news events with terrifying accuracy—posing existential threats to truth in diplomacy and politics.

3. Coordinated Inauthentic Behavior

Troll farms, bot networks, and fake accounts manipulate engagement metrics to manufacture consensus, distort trends, or silence dissent.

4. Hashtag Hijacking and Algorithm Gaming

Information warriors strategically use viral hashtags or pay-to-boost content to dominate public discourse, often drowning out legitimate dialogue.

CASE STUDIES IN MODERN INFORMATION WARFARE

1. Russia's Hybrid War in Ukraine (2014–Present)
Even before physical conflict began, Russia waged a digital war to delegitimize Ukraine's sovereignty, spread confusion in Europe, and divide Western responses.

- Tactics included fake videos of "crisis actors," forged documents, and impersonated journalists.
- Ukraine's counteroffensive relied on rapid digital rebuttals, meme culture, and the charisma of leaders like Zelenskyy on social media.
- The war became one of narrative vs. counter-narrative, with international opinion hanging in the balance.

2. The "Infodemic" of COVID-19
The pandemic wasn't just a public health crisis—it was a narrative crisis. Misleading cures, conspiracy theories, and vaccine disinformation spread virally.

- Competing narratives from China, the U.S., and WHO eroded public trust.

- Telegram groups and fringe influencers became primary sources for millions.
- Diplomats had to engage not just with embassies—but with platforms, influencers, and the digitally vulnerable.

3. Myanmar's Social Media Manipulation

Before and during the Rohingya crisis, Myanmar's military used Facebook to incite ethnic hatred, spread false reports, and dehumanize minorities—turning digital platforms into tools of ethnic cleansing.

- The UN later called Facebook a "key enabler" of the violence.
- It exposed the **real-world harm of unregulated narrative warfare**.

WHY NARRATIVES MATTER IN DIPLOMACY

Diplomacy has always been about narrative: how a country is perceived, how its actions are justified, how its intentions are understood.

But in the digital age:

- Narratives emerge bottom-up, not just top-down.
- The speed of storytelling has outpaced verification.
- Truth competes with "truthiness"—things that feel right, regardless of fact.

Narrative warfare challenges diplomats to:

- Reclaim credibility
- Engage emotionally, not just rationally
- Build **resilience**, not just refutation

DEFENSE AND DETERRENCE IN THE NARRATIVE ARENA

1. Digital Literacy as National Security

States must invest in **citizen education**, teaching people how to spot fake news, question sources, and resist manipulation. A digitally literate population is the first line of defense.

2. Pre-Bunking and Narrative Immunization

Just like vaccines, **preemptive counter-narratives** can help "inoculate" audiences against future disinformation—especially during elections, conflicts, or global events.

3. Platform Accountability and Tech Diplomacy

Governments must engage platforms like X, Meta, YouTube, and TikTok as **diplomatic actors**, pressing for:

- Transparent algorithms
- Swift takedowns of harmful content
- International standards on synthetic media

4. Narrative Coalitions

Alliances must move beyond defense pacts to include information solidarity: rapid narrative response teams, cultural diplomacy exchanges, and support for independent media.

ETHICS AND THE FOG OF INFLUENCE

> When information becomes a weapon, integrity becomes the shield

In the digital age, influence is everywhere. It's in the ads we don't notice, the hashtags we retweet without fact-checking, the headlines tailored to our worldview. Influence is subtle, ambient, embedded in design. It's no longer just what we say—it's **how it's framed, who amplifies it, and what is left out.**

As nations embrace digital diplomacy and strategic communications, a hard question arises: **Where is the ethical line between persuasion and manipulation?**

This is the "fog of influence"—an increasingly murky space where narratives are crafted not to inform, but to condition; not to clarify, but to conquer. It's a space where emotional impact often matters more than empirical truth. And where the distinction between public diplomacy and propaganda becomes dangerously thin.

Let's unpack the ethical dilemmas this fog presents—and what diplomatic integrity looks like in this landscape.

1. **Messaging Ethics: When Is Persuasion Propaganda?** Diplomats are storytellers by design. They explain a country's choices to the world. They frame policies to earn trust. But digital influence campaigns now use psychological tools—micro-targeting, emotional priming, viral repetition—not just to explain reality, but to reshape it.

So when does a message cross the line?

- When it **intentionally misleads**, using half-truths or false urgency.
- When it **erases complexity** to produce outrage or binary thinking.
- When it **demonizes others** to build identity.
- When it **borrows the aesthetics of authenticity**—but masks an agenda.

What's ethical isn't always what's legal. And what's persuasive isn't always what's true. That's the danger—and the responsibility.

2. **Means vs. Ends: Do Good Goals Justify Tactics?** Diplomatic communications often pursue noble ends: promoting peace, defending democracy, building alliances. But when these ends are pursued through viral disinformation, troll farms, or emotional manipulation—what is truly gained?

Is it ethical to:
- Spread unverified stories to galvanize global opinion—even if the underlying cause is just?
- Promote "feel-good" lies for unity, when truth might sow dissent?

- Create fictional enemies to mobilize allies?

These are not hypothetical questions—they are strategic decisions made daily in ministries and campaign war rooms. And they matter. Because every shortcut taken in trust erodes the very foundation upon which diplomacy depends.

3. **Dehumanization Risk: Pawns in Narrative War**
   In information warfare, it's easy to reduce people to symbols:
   - Refugees as "waves"
   - Protesters as "tools of foreign powers"
   - Opponents as "puppets"
   - Victims as "collateral content"

But when people become props in someone else's story, their dignity is lost. Diplomacy must **restore personhood** in a world that too often edits it out.

Ethical narrative strategy begins by asking: *Does this story honor the human beings it depicts—or exploit them for clicks and clout?*

4. **The Ethics of Silence: When Silence Is a Statement.** Not every ethical failure in information

warfare comes from speech. Sometimes, silence is complicity.

- Silence when allies spread hate
- Silence when a state-supported lie goes viral
- Silence when truth is inconvenient for one's geopolitical goals

In a digital world, withholding clarification or truth can be as harmful as active deception. Ethical diplomacy means having the courage to speak even when silence is easier—or more politically comfortable.

5. **Who Holds the Storytellers Accountable?** Unlike military operations, which are subject to formal chains of command and rules of engagement, narrative warfare is often informal, deniable, and diffuse. It's led not just by governments, but by PR firms, influencers, proxy networks, and outsourced digital consultants.

Who is accountable when:

- A manipulated video changes the course of a protest?

- A government pays for a campaign that incites violence abroad?
- A ministry hires an agency to run sockpuppet accounts targeting journalists?

We need **governance models** for digital influence—standards, oversight, and consequences. But we also need something harder to codify: internal diplomatic ethics. The kind that asks not "Can we get away with this?" but "What kind of world does this action help create?"

6. Integrity as Strategic Advantage

There's a false belief in diplomacy that truth is weak—that narrative warfare requires ruthlessness to win.

- But history shows otherwise.
- Truth, when told clearly and consistently, builds lasting credibility.
- Ethical storytelling may not go viral—but it builds resilient relationships.
- Being trusted is **more powerful than being loud**.

In an age where everyone is selling a story, the greatest differentiator is not noise—it is honesty. The most effective diplomats in the future will not be those who manipulate the best. They will be those who are trusted the most.

In Summary: A Compass in the Chaos

Information warfare will not disappear. The tools will only grow more sophisticated. The temptation to cut corners—to win the narrative by any means necessary—will intensify.

But the antidote to manipulation is not more manipulation.

It is **discernment. Accountability. Transparency. Ethics.**

Diplomacy must reclaim these as assets—not as burdens. Because in a world drowning in spin, **integrity is revolutionary**.

Let us ensure that the stories we tell today do not come back tomorrow as regrets—but instead as the foundation of a more truthful, humane, and connected global future.

CONCLUSION: STORYTELLING AS STRATEGY—AND RESPONSIBILITY

In a time when people believe what they feel more than what they read, the **storyteller holds enormous power**. And in today's diplomacy, that storyteller may not be a president—it might be a YouTuber, a deepfake, or a cleverly crafted meme.

Narratives shape perception.

Perception shapes policy.

Policy shapes lives.

As such, diplomats must become guardians of truth in an era that often treats truth as optional. They must lead with clarity in the chaos of contradiction.

And they must remember: in the end, the most effective counter to weaponized disinformation isn't censorship—it's **credible, compassionate, and consistent storytelling** rooted in shared humanity.

This is the real battlefield,

and the pen is still mightier than the sword.

Chapter 6: The Role of Artificial Intelligence in Diplomacy

BEYOND THE ALGORITHM: NAVIGATING POWER, PRECISION, AND PURPOSE IN THE AGE OF AI

Artificial Intelligence has long passed the realm of science fiction and entered the heart of geopolitical reality. Once thought of as a futuristic tool for automation, AI is now a strategic instrument—a quiet architect of influence, insight, and even instability.

AI is not just transforming how we live; it's beginning to transform how we **govern, negotiate, and relate as nations**. And in diplomacy—a field traditionally rooted in nuance, language, and human judgment—AI is emerging not as a replacement, but as a force multiplier.

But as with all power, it comes with risk.

This chapter explores the multiple dimensions of AI in diplomacy: as a tool, a partner, a disruptor, and a new frontier for global cooperation—or competition.

WHERE AI IS ALREADY CHANGING DIPLOMACY

Though still emerging, AI is quietly integrating into core functions of modern diplomacy:

1. Real-Time Information Analysis
AI tools can rapidly process and summarize large volumes of unstructured data—speeches, social media, news reports, leaked documents—giving diplomats **faster situational awareness** in crises.

- Embassies use AI to monitor disinformation in real time.
- Machine learning helps identify patterns in regional instability or electoral interference.

2. Predictive Analytics in Foreign Policy
Governments are beginning to use AI-powered simulations to anticipate:
- Political unrest
- Public sentiment swings

- Climate-driven migration

These forecasts help guide strategic decision-making before crises escalate.

3. Translation and Communication

Natural Language Processing (NLP) has revolutionized cross-linguistic engagement. Real-time translation tools improve accessibility between diplomats and local communities—**breaking linguistic barriers in a field built on dialogue**.

4. AI in Negotiation Support

AI-driven modeling can simulate negotiation outcomes or offer real-time feedback on proposed treaty clauses—providing **scenario analysis** and identifying unintended consequences.

DIPLOMATS AS AI STRATEGISTS— NOT CODERS

The diplomat of the future doesn't need to write code—but they do need to understand **how code shapes power**.

Diplomats must develop AI literacy, including:

- Knowing what AI can and cannot do
- Recognizing algorithmic bias
- Understanding how AI tools may reflect the values—or blind spots—of their creators
- Engaging with ethical frameworks for AI governance

Just as 20th-century diplomats needed to understand oil pipelines and missile treaties, today's diplomats must understand **data flows, neural networks, and machine inference**.

THE ETHICAL EDGE OF AI IN DIPLOMACY

AI introduces extraordinary efficiency—but it also amplifies ethical challenges.

1. Bias and Inequity

AI systems often replicate existing social and geopolitical biases—resulting in discriminatory profiling, misinterpretation of cultural nuance, or exclusion of marginalized voices.

A diplomatic decision made with AI input must always ask: Whose data trained this? **Whose worldview is embedded in the system?**

2. Delegation of Judgment

Can an algorithm understand dignity? Grief? Cultural symbolism? **No.**

Diplomatic judgment requires **human empathy**, contextual reading, and moral deliberation—things AI cannot replicate. Over-relying on AI risks dehumanizing foreign policy decisions.

3. Dual-Use Risks

AI developed for climate forecasting or border control can easily be repurposed for surveillance, suppression, or cyber warfare. Diplomats must **negotiate safeguards**, not just celebrate capability.

CASE STUDIES: AI IN ACTION ACROSS THE GLOBE

1. Estonia: AI-Powered Digital Statecraft

Estonia has integrated AI into digital governance, including public service delivery, digital identity, and even automated policy drafting. Estonian diplomats use AI tools to monitor public discourse and optimize consular workflows.

This is **diplomacy at the speed of code**—transparent, efficient, and citizen-focused.

2. UN: AI for Peacekeeping and Mediation

The UN is exploring AI for:

- Predicting outbreaks of violence in conflict zones
- Detecting online incitement to genocide
- Supporting digital ceasefire monitoring

These applications show how AI can enhance **early warning systems and diplomatic intervention**.

3. China and the Geopolitics of Algorithmic Governance

China's Belt and Road Initiative increasingly includes AI surveillance infrastructure, exported globally. Its AI governance model favors state control over openness, raising diplomatic tensions with democratic states over values and cyber sovereignty.

AI is not just a tool—it is becoming a **diplomatic value proposition**.

GLOBAL AI GOVERNANCE: A DIPLOMATIC IMPERATIVE

As AI systems shape international norms, who sets the standards matters.

- Currently, AI governance is fragmented:
- The EU emphasizes regulation and rights.
- The U.S. leans toward innovation and market freedom.
- China promotes state-centered AI development.

Without a **multilateral framework**, the world risks digital fragmentation and AI "arms races."

What's needed:

- A Global AI Accord—defining ethical principles, red lines, and transparency standards.
- Inclusive AI diplomacy—where small states and civil society have a voice.
- A "UN for AI"—or at least an agile global platform for ongoing coordination.

HUMAN-CENTRIC AI DIPLOMACY: A NEW KIND OF LEADERSHIP

The goal is not to remove humans from diplomacy. It's to empower diplomats with tools that:

- Reveal hidden insights
- Reduce human error
- Improve speed—without sacrificing soul

AI must **serve diplomacy's purpose**, not redefine it. And that purpose is peace, dignity, cooperation, and shared progress.

As such, AI diplomacy must:

- Prioritize **ethics over efficiency**
- Value **inclusion over innovation alone**
- Champion **truth over optimization**

CONCLUSION: BEYOND INTELLIGENCE—TOWARD WISDOM

Artificial Intelligence can predict patterns. It can even mimic tone. But it cannot feel responsibility, nor can it hold itself accountable.

That is the diplomat's burden—and gift.

The future of diplomacy lies in **the partnership between human judgment and machine insight**. But humans must always lead that partnership. AI may inform, but it cannot decide. It may accelerate, but it cannot empathize. It may optimize, but it cannot care.

In the age of AI, diplomacy's true power will lie not in its technology, but in its **discernment, values, and vision**.

Because the future we program today becomes the world we inherit tomorrow. Let us ensure that both the code—and the conduct—reflect the best of what humanity can offer.

Part II: Narrative and Perception

Chapter 7: Crisis Diplomacy in the Digital Age

WHEN THE WORLD WATCHES – AND REACTS – BEFORE THE FACTS ARE CELAR

Crisis used to travel slowly. News of conflict, disaster, or scandal would move through wires, press briefings, and official statements. Diplomats had time to verify, respond, and calibrate. Today, crisis erupts live—on timelines, group chats, livestreams, and message boards. It's instantaneous. It's visual. And often, it's viral before it's understood.

For the modern diplomat, a crisis isn't just about the event itself. It's about the narrative that forms around it, the trust that must be sustained through it, and the tools available to navigate it.

In this chapter, we examine how digital platforms have reshaped the very nature of crisis diplomacy, highlighting strategies, missteps, and the human dimension of leading under digital siege.

THE DIGITAL NATURE OF MODERN CRISES

Crises in the digital age come in many forms:

- Natural disasters amplified by misinformation.
- Cyberattacks that paralyze infrastructure or expose secrets.
- Geopolitical incidents where digital footage spreads faster than fact-checkers.
- Disinformation campaigns that provoke unrest or undermine credibility.

But what unites them is velocity. Within minutes, a crisis can become global. Reputations can falter. Alliances can fray. And once a narrative takes hold online, it can become harder to correct than the event itself.

CHALLENGES FOR CRISIS DIPLOMACY IN A DIGITAL LANDSCAPE

1. Speed vs. Accuracy

Diplomats must respond quickly—but also responsibly. In the digital world, delays are interpreted as weakness or guilt. Yet premature statements can worsen the crisis. Balancing speed and truth is a constant tension.

2. Narrative Fragmentation

A single incident can spawn hundreds of interpretations. Conspiracy theories, manipulated videos, and fake documents can shape public perception before any official account is shared.

3. Loss of Control

Diplomats no longer control the channels of information. Citizens, influencers, bots, and foreign actors all participate in shaping the message.

4. Diplomatic Escalation Online

Public exchanges between state actors on platforms like X (Twitter) or Telegram can replace or upstage formal diplomatic communication. Emotions can overtake strategy.

TOOLS AND TACTICS FOR DIGITAL CRISIS RESPONSE

1. Real-Time Monitoring

Embassies and ministries now rely on dashboards and AI-powered sentiment analysis tools to track how a crisis is unfolding online. This intelligence allows diplomats to anticipate spikes, identify misinformation trends, and adapt messaging quickly.

2. Digital Spokesperson Protocols

Just as militaries train for emergency press briefings, digital diplomacy units prepare templated responses for different types of crises, ensuring that:

- Accurate information is shared fast.
- Key messages are platform-appropriate.
- National values and empathy are visibly communicated.

3. Pre-established Coalitions

Trust cannot be built in the middle of a storm. That's why diplomats increasingly rely on pre-crisis relationships with journalists, tech platforms, NGOs, and even influencers. These networks can help amplify credible information and correct falsehoods.

4. Listening First, Then Leading

Rather than rushing to dominate the narrative, effective digital crisis diplomacy starts with listening—to affected communities, online chatter, and diplomatic peers. This ensures response efforts are relevant, respectful, and rooted in context.

CASE STUDY 1: COVID-19 AND THE INFODEMIC

When Viruses Go Viral: Diplomacy in the Age of Information Contagion

When the world first heard of a mysterious illness spreading in Wuhan, few could have predicted the storm it would unleash. By early 2020, COVID-19 had stopped the world in its tracks. But while leaders scrambled to deploy ventilators and vaccine pipelines, a second pandemic surged—this one not biological, but digital.

The World Health Organization called it an "infodemic"—a tidal wave of information, rumors, conspiracies, and outright disinformation that moved faster than the virus itself. It didn't just spread; it mutated, echoed, and found new hosts every second. And just like the virus, this information contagion didn't respect borders, institutions, or trust.

In this new battlefield, diplomats were no longer just crisis managers or deal-makers—they were narrative responders, digital communicators, and frontline defenders against chaos.

Anatomy of a Digital Plague

The infodemic took shape in layers. First came confusion. Even experts didn't know how the virus spread, how long it lasted, or how best to treat it. Science was evolving by the hour, and in that vacuum, fear crept in.

Then came the geopolitics. Some voices blamed China's Wuhan lab; others pointed fingers at Western powers for weaponizing panic. Accusations flew between states, while platforms like Facebook, Twitter, and Telegram became battlegrounds for competing truths.

Meanwhile, pseudoscience thrived. Entire communities shared videos claiming 5G towers caused COVID. Voice notes on WhatsApp suggested garlic or bleach as cures. Bill Gates became the unlikely villain in a global conspiracy theory. And all the while, trust in public health institutions deteriorated.

Suddenly, wearing a mask was no longer a public health act—it was a political statement. The infodemic had transformed a health crisis into a global identity war.

Diplomacy Reimagined: Combating Lies in Real Time. Faced with this digital wildfire, the usual tools of diplomacy—carefully worded communiqués, official press releases, high-level summits—proved too slow.

The WHO took a frontline role:
- Partnering with tech giants like YouTube and Facebook to flag misleading content.
- Translating weekly briefings into dozens of languages.
- Launching a "Mythbusters" campaign filled with visuals and infographics designed for viral sharing.

Governments tried, with mixed results:
- Singapore created a multilingual, real-time digital dashboard (Gov.sg) to provide facts, debunk rumors, and issue health advisories.
- The UAE tapped influencers and religious leaders to reach diverse communities in targeted, culturally sensitive ways.

- In the U.S., federal messaging faltered. Conflicting statements from agencies like the CDC and White House created confusion rather than clarity—especially in early months.

Diplomatic missions adapted on the fly:
- Embassies became digital lifelines, updating citizens on lockdowns, flights, and health regulations via social media.
- Ministries like the UK's FCDO produced tailored digital content to counter disinformation in diaspora communities.
- Digital diplomacy teams worked overtime—not just to inform, but to calm, connect, and regain trust.

Misinformation as Strategy:
A New Form of Influence

While some states fought disinformation, others leveraged it.

China launched what became known as "mask diplomacy," sending personal protective equipment to over 150 countries, branded with messages like "From China, With Love." State-run media carefully documented every delivery, promoting Beijing's generosity and undermining criticism of its pandemic response.

Russia pushed a different narrative—casting doubt on Western vaccines while promoting Sputnik V as a safer alternative. This strategy found particular traction in Africa and parts of Eastern Europe, where historical skepticism toward the West still lingers.

Meanwhile, troll farms and fake accounts—backed by various states—flooded platforms with conflicting stories, sowing confusion and distrust. Cyberattacks on vaccine research institutions and public health bodies further blurred the line between espionage and narrative warfare.

The pandemic became not just a public health emergency, but a geopolitical stage—where influence was measured in hashtags, likes, and trending videos.

Lessons in Digital Crisis Diplomacy
The infodemic changed the rules.

1. **Speed is everything.** In the digital world, delays are deadly. The first story to go viral often becomes the default belief.

2. **Digital capacity is now a core diplomatic skill.** Ministries and embassies can no longer treat social media as secondary. It is the front line.

3. **Partnerships matter.** Diplomacy now requires collaboration with platforms, influencers, scientists, and civil society.

4. **Empathy outperforms authority.** In times of panic, people look to familiar voices—religious leaders, local health workers, community figures. The most successful campaigns spoke in human, not bureaucratic, language.

Why It Matters

This case is not just about COVID-19. It's about the digital architecture of future crises. Whether the next emergency is biological, environmental, or political, it will unfold in the feeds of billions.

The concept of the "infodemic" has since entered the vocabulary of the UN, NATO, and diplomatic academies. It has sparked new training, protocols, and collaborations between diplomats, tech experts, and behavioral scientists. A new field—crisis narrative diplomacy—is emerging.

Conclusion: Diplomats in the Age of Feeds

The pandemic revealed an uncomfortable truth: truth alone is not enough. It must be fast. It must be human. It must reach people where they already are—on screens, in seconds, from voices they trust.

As the world prepares for future crises, the role of the diplomat must evolve—not only as a negotiator, but as a communicator, a connector, and a guardian of credible information.

Because in the next crisis, the virus might not wait—and neither will the lies.

CASE STUDY 2: UKRAINE INVASION AND REAL-TIME WARTIME DIPLOMACY (2022-)

*A New Era of Wartime Engagement –
Conducted in Pixels and Posts*

When the first missiles struck Kyiv in the early hours of February 24, 2022, the world witnessed not just the beginning of a full-scale military invasion—but the ignition of a parallel conflict in cyberspace. Tanks rolled into Ukrainian territory, but another army was already mobilizing on smartphones, newsfeeds, and encrypted channels.

This was the world's first truly digital war, where diplomacy, defense, and public messaging collided in real time. And at the heart of this moment stood Ukraine—a nation under siege not just by bombs, but by disinformation, cyberattacks, and the pressure to win the world's attention one tweet, one video, one story at a time.

In this war, **legitimacy became as important as logistics, and perception shaped alliances** faster than treaties.

Zelenskyy: The Face of a New Diplomacy
President Volodymyr Zelenskyy, a former comedian and media professional, emerged not only as Ukraine's wartime leader but as a new archetype of 21st-century diplomat. He wore no suit, sat in no formal chamber, and often broadcast from locations marked by rubble and resistance.

His daily videos—delivered in plain clothes, often against the backdrop of a darkened Kyiv—did more than reassure Ukrainians. They electrified audiences around the world.

Zelenskyy's live diplomacy included:

- **Direct appeals to parliaments:** He spoke to the U.S. Congress, the UK Parliament, the EU, Israel's Knesset, and others—each speech carefully crafted with historical parallels and emotional resonance.
- **Tagging world leaders on social media,** calling out companies, NATO, and tech giants with strategic urgency.
- **Speaking like a citizen, not just a president**—inviting people to identify not just with Ukraine's cause, but with its suffering.

His approach humanized geopolitics. It **collapsed the distance between battlefield and boardroom**, between government and the global public.

The Ministry of Digital Transformation: War in the Cloud. While soldiers fought on the ground, a different front was managed by **Ukraine's Ministry of Digital Transformation**, led by Vice Prime Minister Mykhailo Fedorov. Within hours of the invasion, the ministry launched a coordinated digital resistance unlike anything seen before.

1. Crowdsourced Cyber Army

- The "IT Army of Ukraine" mobilized thousands of ethical hackers worldwide.
- Volunteers executed DDoS attacks on Russian state websites, intercepted military communications, and exposed war crimes.
- Telegram channels became command hubs, issuing missions and sharing results in real time.

2. Cryptocurrency Fundraising

- Within a month, Ukraine raised over $60 million in crypto from global donors.
- Funds were transparently tracked on the blockchain and used to purchase equipment, humanitarian aid, and medical supplies.
- Ukraine even launched NFT collections memorializing key moments of resistance—blending art, memory, and fundraising.

3. **Real-Time Digital Diplomacy**
- Fedorov tweeted directly at companies like Apple, Amazon, Google, and Mastercard, demanding action.
- Most responded—either limiting services in Russia or cutting ties altogether.
- Ukraine's government accounts became tools of corporate diplomacy, applying reputational pressure faster than any legal mechanism could.

Narrative Warfare: The Battle for Truth

Russia, with its well-oiled propaganda machine, launched disinformation at industrial scale. But Ukraine countered not with institutional force, but with authenticity.

Tactics included:
- **Memes as weapons:** Ukrainians turned humor into resistance—posting viral videos of farmers towing Russian tanks and elderly women confronting soldiers.
- **Civilian documentation:** Ordinary people filmed attacks, which were verified by NGOs and news outlets, creating a living archive of war crimes.

- **Fact-checking units:** Journalists, civil society, and tech experts created rapid response teams to debunk false Russian claims before they could spread.

Ukraine showed that **being first wasn't enough—being truthful and relatable was what made the message stick.**

Digital Diplomacy That Moves Markets
The ripple effects of Ukraine's digital strategy reached far beyond sympathy.

Global Impacts:
- **Over 1,000 global companies** withdrew or reduced operations in Russia—many after public pressure campaigns.
- **Western governments approved military aid packages**, influenced not just by formal channels but by public sentiment fueled by viral content.
- **Cultural institutions**, from sports federations to arts organizations, imposed symbolic sanctions.
- **The EU banned RT and Sputnik**, citing their role in spreading disinformation—not just media, but as tools of warfare.

The screen had become the new diplomatic stage. And Ukraine controlled the lighting.

Cyber Diplomacy and Hybrid Conflict

This war also redefined the boundaries of sovereignty.

- **Russian cyberattacks** targeted Ukrainian infrastructure, media, and government systems.
- Ukraine, in turn, relied on support from **private tech companies** like Microsoft and Cloudflare, who offered digital shields and cloud backups.
- The war blurred the lines between **nation-states and corporations**, between national security and commercial services.

Tech platforms were no longer neutral—they were now players in the geopolitical arena.

Lessons in Digital Resistance

1. **Authenticity wins hearts faster than perfection.** Zelenskyy's sincerity inspired not just sympathy—but solidarity.

2. **Speed equals credibility.** A tweet in the first hour could matter more than a press conference a day later.

3. **Cloud sovereignty is national sovereignty.** Ukraine's government data survived because it was backed up outside of its borders.

4. **Everyone is a foreign policy actor now.** Influencers, hackers, and digital citizens all helped shape global response.

5. **Platforms are the new frontlines.** Diplomacy now involves CEOs and algorithms as much as ambassadors and treaties.

Why This Case Matters

Ukraine redefined how war and diplomacy intersect. In doing so, it set a new global standard for real-time engagement during conflict. It showed that a **digitally agile state can challenge even the most powerful military force by controlling narrative, truth, and moral clarity.**

The world didn't just hear about Ukraine's war—they saw it, shared it, and felt it in their feeds. And that emotional immediacy shifted policy faster than any cable ever could.

Conclusion: The War That Was Livestreamed

This war proved that diplomacy is no longer confined to embassies or elite rooms. It now unfolds on timelines, in hashtags, on livestreams, and inside comment sections.

The diplomat of today must be part-communicator, part-crisis manager, part-digital native. And the successful foreign policy of tomorrow will depend not just on power—but on **perception, presence, and platforms.**

Ukraine's digital resistance didn't just fight for territory—it fought for truth. And in the screen age, truth—shared and believed—can still be one of the most powerful weapons of all.

CASE STUDY 3:
BEIRUT PORT EXPLOSION (2020)

When Silence and Speculation Collide

It was just after 6 p.m. on August 4, 2020, when the heart of Beirut erupted. A mushroom cloud rose over the city, a thunderclap cracked windows for miles, and within seconds, a shockwave flattened entire districts.

Over 200 people died, more than 6,500 were injured, and 300,000 lost their homes. In one blinding moment, Lebanon's capital was shattered—by the detonation of 2,750 tons of ammonium nitrate stored, neglected and forgotten, in the port.

But as dust settled over the broken streets, a different kind of trauma began to unfold—not physical, but digital. In the absence of reliable information, a tsunami of speculation swept across the internet. Silence from officials was filled not by facts, but by fears. And so began one of the most revealing case studies in crisis-era diplomacy—not because of what was said, but because of what was not.

A Vacuum of Truth

In the first hours after the explosion, there were no clear statements from Lebanese authorities. Conflicting reports filtered in. No one could—or would—explain how such a massive quantity of explosive material had been left in the port for years without adequate oversight.

Into that silence rushed a torrent of misinformation:

- **Foreign sabotage claims** began circulating, blaming Israeli missiles or American intelligence.

- **Conspiracy theories** emerged around Hezbollah's alleged arms caches, suggesting that the explosion was deliberate—or even part of a geopolitical plot.
- **False casualty reports** and fabricated "leaked" documents flooded social media.
- **Voice notes and WhatsApp chains** warned of second attacks, chemical leaks, or curfews that didn't exist.

Videos—some real, others doctored—spread across Facebook, TikTok, and Twitter. News agencies, overwhelmed by the scale of destruction, struggled to verify stories. Traditional media, many already compromised by political bias, became part of the problem.

With institutional trust already brittle in Lebanon, the explosion didn't just demolish infrastructure—it destroyed any remaining faith in official narratives.

Diplomatic Paralysis and Public Pressure

As confusion turned to anger, diplomatic missions found themselves trapped between duty and danger. Many embassies had staff living in the city. Some consulates had suffered structural damage. And all were inundated—by calls from their citizens, by requests from media, and by the Lebanese public desperate for answers.

But few spoke clearly. Statements, where they existed, were vague—limited to expressions of "condolence" and "concern." There was no coordinated diplomatic strategy, no shared messaging, no authoritative voice to push back against the tide of disinformation.

In the absence of trusted leadership, embassies became targets—not of violence, but of rumor. Some were accused of having foreknowledge of the blast. Others were suspected of involvement in murky foreign plots. None had the tools or rapid response strategies to counter these narratives.

What diplomacy lacked was not capacity—but agility.

When Silence Fails: Examples of Response

Amid the overall vacuum, a few diplomatic missions rose to the occasion.

The French Embassy acted quickly and publicly:
- Within hours, it issued **multilingual statements**, including Arabic and French.

- It posted **verified footage and resources** on how to access aid.
- French President Emmanuel Macron **visited Beirut within 48 hours**, delivering live-streamed speeches that emphasized solidarity, accountability, and urgency.

The German Embassy, while less visible on the ground, stood out digitally:

- It launched an **FAQ microsite** addressing popular rumors in three languages.
- Its Twitter team posted **myth-busting threads** debunking false narratives.
- It **partnered with local media and influencers**, ensuring accurate messaging reached broader audiences.

These embassies didn't just communicate—they showed presence. And in the chaos, **presence was power.**

The Digital Diaspora and Civil Society Fill the Void. As official channels faltered, others stepped up. Lebanese around the world mobilized almost instantly:

- Diaspora communities translated government updates into Arabic and other languages, creating **makeshift information hubs.**
- New digital collectives—like the "Elves of Beirut"—formed to **fact-check viral content** and debunk conspiracies.
- Independent media outlets like **Megaphone** and **Daraj** provided real-time updates, investigative reporting, and verified timelines of events.
- Social media influencers became unexpected diplomats—coordinating fundraising, directing volunteers, and demanding international accountability.

In a moment of institutional breakdown, **civil society became the crisis communicator.**

Lessons in Crisis-Era Diplomacy

1. **Speed is credibility.** In a digital crisis, the first official statement can define the entire narrative. Delays create a vacuum that bad actors rush to fill.

2. **Silence is not neutral.** When embassies or governments stay quiet, they're not invisible—they

become suspicious. The public will fill gaps with speculation.

3. **Local voices matter.** Partnership with journalists, influencers, and community leaders is not optional. These actors often have more reach—and more trust—than formal diplomatic channels.

4. **Pre-crisis preparation is key.** Diplomatic missions must have crisis playbooks: translated messaging templates, digital coordination plans, and platform-specific response strategies.

5. **Transparency builds trust.** Acknowledging what is unknown is often more powerful than offering false certainty. Emotional honesty can ground a chaotic moment.

Why This Case Matters

The Beirut port explosion is not just a case of government failure—it's a case of **narrative collapse**. And it shows what happens when diplomacy doesn't adapt quickly enough to the speed of crisis-driven misinformation.

For diplomatic actors, the lesson is sobering: **it's not enough to be present after the fact**. In today's world, **credibility is won—or lost—in real time.**

Those who spoke, who connected, who offered help with clarity and compassion earned public trust. Those who waited became ghosts in the digital fog.

Conclusion: Presence as Strategy

In the modern age, diplomacy no longer happens solely behind closed doors. It unfolds in real-time, in feeds, and in the minds of the public. The Beirut blast proved that tragedy doesn't wait for policy—and neither does the search for truth.

To be effective in such moments, diplomats must be more than representatives. They must be narrative leaders, empathic communicators, and rapid responders.

The next catastrophe may strike elsewhere. But the question will remain the same:

Who will speak first—and who will be believed?

THE DIPLOMAT AS DIGITAL FIRST RESPONDER

Crisis diplomacy in the digital age is not a theoretical exercise—it's survival. It demands agility, emotional intelligence, communication fluency, and above all, preparedness.

The modern diplomat must be part emergency responder, part narrative strategist, and part digital analyst. And in many cases, the diplomat will be the first voice heard by the international community—and the last hope for clarity amid chaos.

To lead effectively through crisis, diplomats must:

- Communicate swiftly and with moral clarity.
- Build trusted channels long before they're needed.
- Understand the psychological landscape of online audiences.
- Balance discretion with transparency.

Above all, they must stay human in the storm—offering empathy, truth, and connection when the world feels most fragile.

In the next chapter, we explore the ethical dilemmas and governance questions raised by this transformation. As diplomacy merges with digital power, what does it mean to act responsibly? Who sets the rules? And what are the risks of failing to?

CONCLUSION: CALM IN THE EYE OF THE DIGITAL STORM

In times of crisis, diplomacy has always served as the anchor in turbulent seas—offering clarity, building bridges, and preserving dialogue when the world teeters on the edge of rupture. But today's crises are no longer confined to closed-door negotiations or official communiqués. They unfold on screens, in real time, amplified by algorithms and shaped by the emotions of a global audience.

In this hyperconnected world, the role of the diplomat has shifted from background negotiator to front-line communicator. A crisis that once took days to reach the public now unfolds within minutes. The first response is not always a government statement—it may be a tweet, a video, a viral rumor. In this new environment, the tools of diplomacy include dashboards, digital literacy, and narrative agility.

The architecture of crisis has changed. Facts compete with falsehoods. Public sentiment solidifies before official facts emerge. Misinformation—unfiltered and unchecked—can reshape policy environments, spark unrest, or undermine trust in institutions.

And yet, amidst this disruption, the core mission of crisis diplomacy endures: to bring order to chaos, meaning to confusion, and compassion to pain. What has evolved is the medium; what remains is the imperative.

The case studies presented in this chapter—from the global "infodemic" of COVID-19, to the digital defiance of Ukraine, to the narrative vacuum following the Beirut explosion—highlight one essential truth: digital crisis diplomacy is no longer optional. It is central to international engagement, and must be designed, rehearsed, and embedded into every layer of foreign policy strategy.

And critically, even in a digital landscape, the most effective instruments of crisis diplomacy remain deeply human:

- The restraint to verify before reacting.

- The emotional intelligence to read not just the headlines, but the heart of a population.
- The discernment to know when to speak—and when silence holds more meaning.
- The trust built not in the moment of crisis, but in the calm that preceded it.

In an era where crises are viral, borderless, and fast-moving, diplomacy must be equally agile. But agility must not come at the cost of ethics or empathy. It must be grounded in purpose, guided by principle, and executed with care.

In the next chapter, we turn to the ethical responsibilities that accompany this new era of diplomacy—because influence without integrity is not leadership. And in the digital age, the question is no longer whether we engage, but how we choose to lead in shaping the global narratives of tomorrow.

Chapter 8: Cybersecurity Challenges and Responses

IN THE DIGITAL ERA, NATIONAL SECURITY BEGINS AT THE KEYBOARD

Gone are the days when security meant troops on a border or ships in a harbor. Today, a simple phishing email can compromise a government agency. A line of malicious code can black out an entire city. And a well-timed hack can change the course of an election.

In the 21st century, cybersecurity is not a peripheral issue—it is a defining frontier of global security and diplomacy. From state-sponsored cyberattacks to critical infrastructure vulnerabilities, from ransomware groups to digital espionage, the threats are invisible but deeply real.

This chapter explores the evolving landscape of cyber threats, the strategic responses by governments and multilateral institutions, and the urgent need for diplomatic frameworks that protect both people and systems.

THE EXPANDING CYBER THREAT LANDSCAPE

The digital transformation of our economies, governments, and societies has created vast surfaces for attack—and many of them remain poorly defended.

1. Critical Infrastructure at Risk

- Power grids, water treatment plants, hospitals, air traffic systems—all are now connected to digital networks.
- Cyberattacks on these systems can cause mass disruption without a single shot being fired.

2. Electoral Systems and Democratic Integrity

From disinformation campaigns to direct breaches of electoral infrastructure, cyber tools are being used to undermine trust in democracy itself.

3. The Rise of Ransomware as a Business Model
Cybercriminals have turned hacking into a global industry, demanding millions in cryptocurrency from hospitals, schools, and governments.

4. Espionage Reimagined
Cyber espionage enables silent infiltration of foreign ministries, research institutions, and global corporations—with attribution often delayed or denied.

NATION-STATE ACTORS AND THE BLURRED LINES OF WARFARE

What makes cybersecurity uniquely complex is the blurring of lines between state and non-state actors, between espionage and warfare, between criminal and geopolitical motives.

1. Attribution Challenges
Unlike traditional warfare, it's often unclear who is attacking, why, and on whose behalf. Sophisticated attacks can be masked to appear as criminal acts or outsourced to proxy hackers.

2. Hybrid Warfare

Cyber tools are now embedded in broader military strategies:

- Disrupting communication before a kinetic strike
- Spreading panic during political transitions
- Undermining international alliances via leaked documents

Cyber is no longer a separate domain—it is a force multiplier in every conflict.

CYBER NORMS: THE MISSING ARCHITECTURE

Despite the scale of the threat, there is no universally accepted framework for cyber conflict akin to the Geneva Conventions or arms control treaties.

Current Efforts Include:

- UN GGE and OEWG negotiations on responsible state behavior in cyberspace
- The Paris Call for Trust and Security in Cyberspace, a voluntary agreement supported by over 1,000 entities
- Bilateral cyber agreements between major powers

Yet many of these initiatives lack enforcement power or fail to include key voices from the Global South and civil society.

What's Needed:

A global cyber compact that defines:

- Protected digital infrastructure (like hospitals)
- Prohibited cyber actions (e.g., attacks on humanitarian networks)
- Shared incident response protocols

DIGITAL DIPLOMACY IN THE CYBER AGE

Diplomats are no longer just managing geopolitical alliances—they are increasingly becoming cyber crisis responders, trust-builders, and norm-setters.

Key Roles for Diplomats Include:

- **De-escalating cyber incidents** through backchannels and public signaling
- **Coordinating joint responses** to global threats like WannaCry or NotPetya

- **Advocating for human rights online**, particularly in surveillance and censorship debates
- **Engaging with tech companies**, who often detect breaches before governments do

Diplomatic missions now require cyber attachés and technical advisors alongside political experts. It's a team effort—and it must be resourced accordingly.

CASE STUDIES: LESSONS IN CYBER DIPLOMACY

1. **Estonia: From Crisis to Cyber Superpower.** After a crippling cyberattack in 2007—believed to be linked to geopolitical tensions with Russia—Estonia rebuilt its digital infrastructure with resilience and transparency at the core.
- Today, Estonia is home to NATO's Cooperative Cyber Defence Centre of Excellence.
- It serves as a model for public-private cyber collaboration and digital governance.

2. **SolarWinds and the Global Wake-Up Call.** The 2020 SolarWinds hack, attributed to a state-backed actor, affected thousands of U.S. federal agencies and private companies.

- It revealed vulnerabilities not in hardware, but in software supply chains.
- The response involved sanctions, international condemnation, and new security protocols—marking a shift from covert tolerance to overt retaliation.

3. **Cyber Peace Institute.** This Geneva-based organization advocates for cyber accountability, particularly when civilians are harmed.
- It pushes for digital conflict norms like humanitarian law.
- Its work underscores that cyber warfare is not abstract—it affects real lives, in real time.

A HUMAN-CENTERED APPROACH TO CYBERSECURITY

While cybersecurity is often framed in technical terms, its human consequences are profound:

- Patients turned away from hacked hospitals
- Journalists surveilled or silenced
- Refugees denied digital access due to state censorship

We must view cybersecurity not just as infrastructure protection—but as human security.

A Human-Centered Approach Must:

- Prioritize the protection of vulnerable populations
- Respect privacy and digital dignity
- Include ethical considerations in national cyber strategies
- Ensure equity in digital defense capacity, especially for developing nations

Cybersecurity is not just a technical challenge—it is a moral and diplomatic one.

CONCLUSION:
TOWARD A SAFER DIGITAL FUTURE

Cybersecurity is no longer a niche issue for IT departments or intelligence agencies. It is a cornerstone of **national sovereignty, international peace, and global diplomacy.**

The threats will grow. The tools will evolve. And the conflicts will become more subtle, more sophisticated, and more entangled with everyday life.

But we are not powerless.

With the right norms, partnerships, and human-centered policies, we can build a digital future that is not only **connected—but secure, trusted, and just.**

This is the new frontier of diplomacy:

Where defense means dialogue,

Where security means cooperation,

And where peace is no longer just the absence of war—but the **presence of digital resilience.**

Chapter 9: Crisis Management and Communication in the Digital Era

"In a time of chaos, silence is not neutrality – it is failure. In a digital world, crisis leadership begins with showing up, listening deeply, and speaking clearly."

INTRODUCTION: WHEN CRISES GO VIRAL BEFORE THEY ARE UNDERSTOOD

The anatomy of crisis has changed. Once, a government could learn of a disaster, assemble its experts, and prepare a response before addressing the public. Today, by the time a foreign ministry receives a situation report, the world may have already seen it live on a TikTok stream. A whistleblower halfway across the globe might have already leaked the images. Public outrage might be trending under a hashtag created by a teenager with no diplomatic credentials but a viral message.

Crisis management in the digital era is not merely a logistical function—it is a real-time, public, and deeply human endeavor. It is defined by speed, visibility, and the power of perception. Those in positions of leadership no longer control when or how a crisis is first revealed. But they do control whether their response builds trust—or erodes it.

This chapter explores how governments, diplomats, and civil society organizations must adapt to this new reality, where **authenticity is currency, narrative is strategy**, and **digital agility is survival**.

THE HUMAN FACE OF DIGITAL CRISIS RESPONSE

In times of crisis, people do not look to institutions—they look to people. A trembling voice in a livestream. A mother filming her child's displacement. A nurse showing an overcrowded ICU. These human moments break through the noise in ways press statements rarely do.

And so, leadership in a digital crisis demands more than protocols and press conferences. It demands humanity, presence, and an unflinching willingness to speak—imperfectly if necessary—rather than say nothing at all.

We have seen time and again that in the absence of rapid, transparent communication, the digital vacuum fills itself—with rumor, rage, or repression. The narrative slips from institutional hands and is reclaimed by those on the ground, with or without government support. This shift is not inherently bad—it is, in fact, democratizing. But it requires governments to earn trust, not expect it.

Crisis communication today is no longer a single, top-down stream. It is a constellation of actors—official and unofficial—shaping public understanding in real time. Diplomats, ministers, local responders, influencers, journalists, bots, AI algorithms, and ordinary citizens all contribute to how a crisis is perceived and managed. This complexity cannot be controlled—but it can be navigated.

CASE STUDY 1: UKRAINE'S DIGITAL DEFIANCE AND NARRATIVE LEADERSHIP

A War of Missiles and Messages

When Russia launched its full-scale invasion of Ukraine on February 24, 2022, the world braced for tanks, airstrikes, and territorial collapse. What it didn't anticipate was a very different kind of resistance—broadcast live, pixel by pixel, from smartphones and encrypted apps. In parallel to the military defense of cities and borders, Ukraine waged a second war—one for global perception, moral clarity, and digital sovereignty.

At the center of this resistance was President Volodymyr Zelenskyy—a former comedian turned wartime leader—who emerged not as a traditional statesman, but as the digital voice of a nation under siege. Shedding the presidential suit for a green military shirt, Zelenskyy filmed himself alongside his cabinet in Kyiv's war-torn streets. "We are all here," he declared in a shaky handheld video that traveled around the world. In that moment, Ukraine became more than a country—it became a story.

Zelenskyy didn't follow the script of diplomacy. He bypassed foreign ministries and spoke directly to citizens abroad, addressing parliaments from his iPhone, quoting Shakespeare in London and invoking Martin Luther King Jr. in Washington. His words weren't polished—they were raw, urgent, and tailored for each audience. They reminded the world not just of war, but of shared humanity. He didn't only ask for weapons. He asked for solidarity.

But Ukraine's digital defiance wasn't just about leadership at the top. Behind the scenes, the Ministry of Digital Transformation was building an entirely new kind of war room—one powered by cloud storage, encrypted messaging, and civic tech. They mobilized the "IT Army of Ukraine," a volunteer cyber force of global hackers and tech experts who disrupted Russian propaganda networks, targeted military databases, and fought disinformation from behind screens.

The government's Diia app, originally designed for civil services, was quickly adapted for wartime use. Citizens used it to report enemy troop movements, access public alerts, and retain digital IDs even when displaced. Meanwhile, Ukraine began accepting—and tracking—donations in cryptocurrency, raising over

$100 million in just weeks. Blockchain records ensured transparency, and NFTs were minted as both digital war memorials and tools for fundraising. Ukraine became the first nation to conduct a portion of its defense on-chain.

Online, humor became a weapon of its own. Memes mocked the Russian advance, portraying grandmothers with pickle jars as anti-tank warriors and Russian soldiers asking for directions in villages that no longer existed. Telegram channels like "Ukraine Now" coordinated citizen reporting, while open-source intelligence groups documented war crimes in real time—preserving truth for future tribunals.

Ukraine's counter-disinformation tactics were fast, layered, and deeply human. The government partnered with NGOs to verify videos, geolocate images, and expose fake narratives, deepfakes, and staged attacks. When Russia tried to flood the information space with confusion, Ukraine responded with transparency, emotion, and speed.

What made this digital strategy revolutionary was not just its innovation, but its inclusivity. Diaspora Ukrainians ran global media campaigns. Influencers amplified Zelenskyy's pleas. Citizens became journalists, tech developers, and frontline narrators. Everyone had a role, and every voice mattered.

The global ripple effects were undeniable. Hashtags turned into headlines. Viral videos of resistance spurred parliaments to act. Within weeks, countries across Europe and North America sent aid, opened borders to refugees, and imposed sweeping sanctions on Russia. Over 1,000 companies pulled out of the Russian market—many after being publicly called out by Ukrainian digital activists.

Sports federations banned Russian athletes. Film festivals disinvited Russian entries. Ukrainian culture itself became an act of resistance—songs, dances, embroidery, and language amplified across Instagram, TikTok, and YouTube as symbols of defiance.

Why This Case Matters
Ukraine's digital war was not an accident. It was a deliberate, coordinated strategy rooted in truth, urgency, and emotional connection. It showed that in a world of information overload, authenticity cuts through the noise. When done right, narrative diplomacy isn't a side campaign—it's the frontline.

This case marked a turning point in how governments engage the public during conflict. Small and medium nations, previously drowned out by global powers, now had tools to speak—and be heard. Ukraine set a blueprint for digital resistance that has already begun to shape how other countries prepare for future crises.

Lessons Learned

1. **Narrative is national security.** Controlling your story can be as vital as defending your soil.

2. **Authenticity beats perfection.** A handheld video with emotion resonates more than a polished speech without soul.

3. **Government must be a platform.** Ukraine empowered not just ministries, but citizens, tech firms, NGOs, and diasporas to participate.

4. **Digital participation scales trust.** By involving everyone, from meme creators to coders, Ukraine decentralized resilience.

5. **Crisis communication is now real-time diplomacy.** Waiting days—or even hours—to respond risks ceding control of the truth.

Conclusion: The New Rules of Engagement

Ukraine didn't just resist invasion—it redefined leadership in the digital age. President Zelenskyy became a symbol not through spin, but through sincerity. The Ministry of Digital Transformation proved that code could be as powerful as steel. And the Ukrainian people reminded the world that courage isn't always shown on the battlefield—it's also uploaded, streamed, and shared.

In the 21st century, survival is not only about strength. It's about story. Ukraine taught us that digital infrastructure, emotional intelligence, and narrative clarity are no longer luxuries of peacetime—they are weapons of sovereignty.

The war for Ukraine continues. But the war for truth? In many ways, Ukraine already won.

CASE STUDY 2: TAIWAN'S TRANSPARENT PANDEMIC COMMUNICATION

Trust in Crisis: How Digital Democracy Endured

In the final days of 2019, while most of the world welcomed a new year with celebrations, something quieter stirred across the Taiwan Strait. News emerged of an unusual illness spreading in Wuhan. For many governments, it seemed distant. But for Taiwan—a densely populated island with strong ties to China and painful memories of the 2003 SARS outbreak—it felt dangerously close.

Taiwan moved fast. Airport checks were instituted on New Year's Eve. By January 20, 2020, the government activated its Central Epidemic Command Center (CECC), a cross-ministerial crisis body. It wasn't the speed alone that made Taiwan exceptional. It was the philosophy behind its response: a deep, institutional belief that public trust—not fear, not force—would be its strongest defense.

From the outset, the government made a rare promise: full transparency. Officials spoke daily to the nation in livestreamed briefings, where they didn't dodge hard questions or sugarcoat numbers. The tone was calm, factual, and human. Mistakes were acknowledged. Corrections were made publicly. There was no room for rumor to outrun truth.

Central to this approach was Audrey Tang, Taiwan's unconventional Digital Minister. A former hacker and open-source evangelist, Tang understood that information—when shared openly and creatively—could itself be a vaccine against fear. Under her leadership, Taiwan released real-time data on everything from mask supplies to clinic locations. When mask shortages sparked anxiety, the government didn't issue defensive press releases. It shared its supply chain data with the public. Within days, civic technologists across the country had built apps and maps to track mask availability by neighborhood.

It wasn't just about openness—it was about participation. Citizens became co-creators of solutions. Tang's team ran an online suggestion platform where people submitted ideas for pandemic response. Some were implemented within days. Youth volunteers helped elderly citizens sign up for vaccines. Religious leaders translated health messages into indigenous languages. Civic tech groups like g0v fact-checked misinformation in real time.

Even the tone of Taiwan's crisis messaging reflected care. A cartoon Shiba Inu named Zongchai became a national icon, gently reminding citizens about social distancing and mask-wearing. When fake news spread on chat apps like LINE, the government responded not with punishment, but with verified infographics that were clear, visual, and often laced with humor. Audrey Tang's guiding principle was simple: "Humor over rumor. Clarity over confusion. Participation over propaganda."

By the end of 2020, Taiwan had recorded fewer than 1,000 cases and fewer than 10 deaths—without the need for mass lockdowns or police-enforced curfews. The economy stayed largely open. Schools remained in session. And perhaps most importantly, the public never lost trust in its institutions.

Why This Case Matters
Taiwan's story is not just a case of public health success—it's a powerful model of ethical crisis leadership in the digital age. In a time when many governments responded with censorship, coercion, or denial, Taiwan demonstrated that the foundation of resilience is trust.

This case illustrates how small democracies, often overlooked in global decision-making bodies, can lead not by size or wealth, but by example. In a geopolitical landscape where Taiwan remains diplomatically isolated—barred from the World Health

Organization—it nonetheless emerged as a source of guidance, innovation, and moral clarity.

Lessons Learned

1. **Transparency is not weakness—it's strength.** Taiwan proved that open information doesn't cause panic; it builds confidence.

2. **Speed matters, but trust matters more.** Rapid decisions were matched with equally rapid communication. The government earned trust in real time.

3. **Empathy is strategic.** From friendly mascots to multilingual outreach, Taiwan treated its people with respect and warmth.

4. **Participation scales capacity.** The government didn't act alone. It enabled a whole-of-society response—volunteers, developers, educators, and ordinary citizens.

5. **Digital infrastructure must be democratic.** Technology wasn't used to surveil or control—it was used to empower and connect.

Global Impact

Taiwan's response earned praise from journalists, public health experts, and digital rights advocates worldwide. Countries with far greater resources looked to Taiwan for best practices in data sharing, rumor control, and citizen engagement.

Despite its exclusion from WHO proceedings due to political pressure from China, Taiwan's success forced the global community to take notice. In many ways, its pandemic diplomacy became a new form of soft power. Governments reached out not just for masks, but for advice.

Taiwan's leadership wasn't loud—but it was clear. In an age of noise, clarity is powerful.

Conclusion: Democracy as the Ultimate Defense

Taiwan's pandemic response wasn't perfect. There were challenges, missteps, and learning curves. But the core lesson endures: in a time of fear, people look to those who tell them the truth.

Taiwan didn't lock people in their homes. It invited them into the process. It didn't issue blind commands. It told stories, shared tools, and asked for help. And the public responded—not with resistance, but with solidarity.

In an era where misinformation spreads faster than viruses, and where many democracies struggle to maintain public faith, Taiwan's case offers a rare thing: hope. It shows that with transparency, empathy, and digital cooperation, even a small island can stand tall in a global storm.

It reminds us that trust is not something you demand. It's something you build—message by message, day by day.

CASE STUDY 3: BEIRUT'S SILENCE AND THE PRICE OF MISTRUST

When the Absence of Truth
Becomes a Political Actor

On the afternoon of August 4, 2020, the heart of Beirut was shattered in seconds. A colossal explosion tore through the port, leveling entire neighborhoods and sending a mushroom cloud into the sky. It was one of the largest non-nuclear blasts in history—over 200 lives lost, 6,500 injured, and 300,000 displaced in a city already on the brink of collapse. But in the hours and days that followed, another rupture emerged—one less visible, yet equally devastating: silence.

The Lebanese government, long mired in dysfunction and public distrust, failed to respond with clarity, unity, or even basic coordination. Ministries contradicted each other. Statements were delayed or incomplete. No digital dashboard emerged. No senior leader took the digital stage to console, explain, or lead. In the critical vacuum left by this absence, confusion metastasized into anger, and misinformation filled every gap.

Instead of answers, the Lebanese people received rumors. Was it an Israeli strike? A Hezbollah weapons cache? A government cover-up? On WhatsApp and Twitter, doctored videos circulated. Social media was flooded with fake casualty lists and blame-shifting conspiracies. Foreign embassies were drawn into the maelstrom—accused, without basis, of prior knowledge or even complicity. With no central voice of trust, everything became believable. Or unbelievable. Truth lost its anchor.

For diplomats in Beirut, the blast posed an operational and reputational nightmare. Embassies near the port were damaged. Staff were injured. Protests swelled near diplomatic quarters. Yet their communications were, for the most part, formulaic: short statements of condolence, vague promises of support. Few addressed the storm of misinformation directly. There was no multilateral effort to shape a coherent narrative. Each mission acted alone—and in that solitude, they missed an opportunity to lead.

But the vacuum did not remain empty. Civil society stepped forward—scraped, grieving, but determined. Independent media platforms like Megaphone and Daraj built verified timelines, debunked false reports, and named the officials responsible for years of neglect. Civic technologists used WhatsApp, Instagram, and open-source tools to locate the missing, track damage zones, and coordinate food and medical aid. Youth volunteers swept glass from the streets and posted live updates, while the Lebanese diaspora launched digital fundraisers, crisis maps, and rumor-checking groups like "Elves of Beirut."

One of the few bright lights in the diplomatic sphere came from France. Within days, President Emmanuel Macron visited Beirut, walking among the rubble and grieving families. His speech was livestreamed. His presence felt. It wasn't flawless diplomacy—but it was visible, empathetic, and real. The German Embassy followed suit with an Arabic-language FAQ debunking viral rumors and explaining their aid efforts. These were small gestures in a sea of uncertainty, but they mattered.

Why This Case Matters

Beirut's tragedy is a stark reminder that silence is not neutral. In the digital age, the absence of official communication is itself a powerful force—one that breeds suspicion, conspiracy, and collapse. Governments and diplomats must realize: when people search for answers and find none, they will fill the silence with fear.

Lessons Learned

1. **Crisis leadership must be digital and immediate.** The first hours after disaster are not just logistical—they are narrative-defining.

2. **Embassies are not invisible.** Even when not at fault, foreign missions are part of the public's emotional and political landscape.

3. **Coordination is credibility.** Fragmented responses weaken trust. Unified messaging—even across multiple actors—strengthens it.

4. **Empathy is non-negotiable.** Real-time emotion—grief, compassion, solidarity—must accompany real-time facts.

5. **Preparedness includes narratives.** Diplomatic kits should include rumor response templates, multilingual FAQs, verified image banks, and crisis-tested social media teams.

Global Impact

The Beirut explosion exposed the fragility of trust in post-crisis governance. International observers saw firsthand the costs of narrative neglect. In subsequent crises—natural disasters, armed conflicts, health emergencies—many governments began rethinking their information protocols. Training modules on "strategic empathy," "narrative coordination," and "digital rumor containment" became standard in foreign service academies across Europe and North America.

The case also deepened global conversations around diplomatic responsibilities in fractured states. It raised an uncomfortable truth: sometimes, embassies must speak not just for their countries—but for truth itself.

Conclusion: Filling the Silence with Credibility

Beirut did not lack courage. Its people ran toward the smoke, not away from it. They rebuilt homes, shared resources, and documented truth with phones and tears. What Beirut lacked was leadership—narrative leadership.

In a city crushed by explosion and corruption, citizens became their own diplomats. And the world saw the cost of governmental silence: delayed justice, prolonged trauma, and reputational damage that no amount of aid can reverse.

The Beirut case teaches us this: communication is not a luxury in crisis. It is infrastructure. It is protection. And in a world where tragedies unfold in real time, credibility is the most valuable form of aid a government—or an embassy—can offer.

CRISIS LEADERSHIP IN A NETWORKED WORLD

These case studies—Ukraine, Taiwan, and Lebanon—reveal a fundamental shift: crisis management today is as much about narrative clarity as it is about logistical coordination. It is about listening as much as leading. It is about meeting fear and confusion with transparency, not distance.

To lead effectively in the digital era, public figures must:

- Communicate **honestly and rapidly**, even when information is incomplete.
- Engage citizens **where they already are**—on the platforms they use, in the language they understand.
- Allow for **two-way dialogue**, not just one-way broadcasts.
- Share data **openly** but contextualize it with compassion and story.

Digital crisis communication is no longer just the domain of press officers. It is core to national resilience. It is diplomacy, public health, disaster response, conflict prevention, and governance—

interwoven into one urgent stream of human experience.

CONCLUSION: WHEN SILENCE SPEAKS LOUDER THAN WORDS

The tragedy of the Beirut port explosion was not only one of physical devastation—it was a crisis of narrative. In a moment when leadership was needed most, the Lebanese state failed to provide even the most basic forms of clarity, compassion, or communication. That failure did not just delay aid or frustrate citizens. It eroded what little trust remained in public institutions and created a vacuum filled by speculation, outrage, and disinformation.

In the digital age, silence is no longer a neutral position. It is a communicative act—one that signals indifference, negligence, or even complicity. The absence of a timely and coordinated narrative after Beirut's disaster allowed falsehoods to flourish and foreign actors to be scapegoated. It turned embassies into targets of suspicion, not because of what they did, but because of what they failed to say.

But this chapter also highlights the power of voices from the margins. Where governments faltered, civil society, independent media, and diaspora communities stepped in. They did not wait for permission. They did not rely on hierarchy. They acted—with digital tools, local knowledge, and emotional intelligence. Their response was not perfect, but it was human, immediate, and effective.

For diplomats, the lessons are clear and urgent:

- Crisis communication must be preemptive, not reactive.
- Narrative leadership is not a luxury—it is a necessity.
- Silence leaves space for speculation. Voice creates space for trust.

Beirut was a warning. In today's information environment, a microphone is always live, a camera always rolling, and an audience always watching. The question is not whether someone will shape the story—but who, how, and to what end.

Diplomacy in the digital age is not just about speaking for your nation—it is about standing with others in their darkest hours. When catastrophe strikes, the most powerful tools at our disposal are not only aid packages or press releases, but authenticity, presence, and truth. These are the currencies of modern credibility.

The Beirut explosion reminds us that when governments go silent, others must speak. And if diplomacy is to remain relevant in the age of digital grief and outrage, it must learn to speak not only on behalf of states, but on behalf of humanity.

Chapter 10: The Geopolitics of Data and Sovereignty

"In the 21st century, sovereignty is measured not just by territory – but by who controls the flow of information."

INTRODUCTION: REDEFINING SOVEREIGNTY IN THE AGE OF THE CLOUD

In a world once governed by territorial maps, power was counted in miles of land and numbers of troops. But in the 21st century, where communication is instantaneous and memory is measured in terabytes, a quiet shift has occurred. The essence of sovereignty—the very idea of what it means for a state to be independent and self-determining—has begun to migrate from the physical to the digital.

Today, data has emerged as one of the most powerful geopolitical currencies. The trails we leave online—our health records, financial transactions, search histories, political opinions, biometric identifiers—are no longer mere byproducts of modern life. They are the raw material of influence, governance, profit, and power. And just like oil or water before them, these invisible resources are sparking a race for control.

Nations now compete not only for natural resources but for digital territory: the infrastructure, legislation, and technical capacity to determine where data lives, how it moves, and who gets to access it. It is a struggle waged quietly through trade agreements, cloud server installations, cybersecurity law, and the tug-of-war over regulatory standards. But its consequences are anything but abstract.

At the heart of this new frontier lies a profound question: **can sovereignty survive in a borderless digital world?** And if so, whose version of sovereignty will prevail?

DIGITAL SOVEREIGNTY: BETWEEN CONTROL AND FREEDOM

The concept of "digital sovereignty" is often defined in policy circles as a nation's ability to govern its own digital infrastructure, data flows, and technological development. But beneath this technical definition lie deeper tensions.

For some governments, digital sovereignty means self-reliance: the right to store data domestically, to regulate foreign technology firms, to defend against cyber intrusions, and to set their own terms for data privacy and digital trade. For others, it becomes a justification for control—over speech, surveillance, and information access.

And for citizens, digital sovereignty is not just an issue of abstract governance—it is a question of who sees them, who profiles them, and who decides what they can or cannot know about the world around them.

This is not a debate reserved for Silicon Valley boardrooms or Brussels policy papers. It is being played out in villages where internet cables are just arriving, in refugee camps connected by biometric ID systems, and in every home where smart devices listen quietly, learning the rhythms of life.

CASE STUDY 1: THE EUROPEAN UNION AND THE ARCHITECTURE OF TRUST

How GDPR Became a Digital Moral Compass

In the spring of 2018, as the digital world roared ahead—fueled by data, driven by algorithms, and ruled by tech giants—a quiet revolution emerged from Brussels. The European Union's General Data Protection Regulation (GDPR) was not announced with fanfare. But its impact would ripple across continents, shifting the terms of debate about the internet, power, and rights in the 21st century.

At its core, the GDPR did something profound: it declared that personal data belongs not to governments, corporations, or platforms—but to people. In an age when data had become currency, the EU insisted it was still a part of human dignity. It was less a law, more a philosophy—a line drawn in digital sand.

Unlike many regulatory frameworks, GDPR wasn't just about legal compliance; it was about values. It was Europe's answer to the growing fear that our lives were being quietly turned into a product. That fear wasn't unfounded. By the late 2010s, surveillance capitalism—an economic model built on the extraction of behavioral data—had become the norm. Targeted ads, predictive policing, opaque algorithms, and invasive tracking had outpaced regulation everywhere. Everywhere, that is, except the EU.

Europe as a Global Rulemaker

What made the GDPR truly transformative wasn't its content alone, but its scope. With the EU's vast consumer market as leverage, any company—whether based in San Francisco, São Paulo, or Singapore—that wanted access to European users had to comply. This regulatory reach effectively turned Brussels into a global digital policymaker.

- Major platforms rewrote their privacy policies.
- Startups hired data protection officers.
- "GDPR compliance" became a line item in corporate strategy decks from Tokyo to Toronto.

This wasn't regulation through force—it was governance through gravity. In a fragmented digital world, the GDPR became a kind of ethical anchor. And in doing so, the EU introduced a new form of soft power: legal diplomacy through values.

The Imperfect Protector

Of course, no legislation is flawless. The GDPR faced criticism on several fronts:

- **Complexity and cost** overwhelmed many small businesses.
- **Loopholes** were exploited by larger players.
- **Enforcement** remained inconsistent, with wide variance between member states.

Still, the core principle endured: individual rights must survive technological disruption. And in that survival lay the EU's long-term strategy—not just to protect its citizens, but to shape the global debate.

What the EU offered through the GDPR was not just law, but leadership. As the United States lagged behind in federal data regulation, and China moved toward total digital centralization, Europe carved out a middle path—a democratic framework for digital dignity.

Balancing Innovation and Responsibility

The GDPR did not arrive without tension. Critics warned that strict data controls would stifle innovation, especially in AI and machine learning. Some startups even relocated to jurisdictions with looser standards. But defenders of GDPR argued that innovation without ethical constraint leads to long-term harm—from biased algorithms to unaccountable surveillance.

This tension pushed the EU to evolve further. In the years following the GDPR, Europe introduced new legislation like:

- **The Digital Services Act (DSA):** holding platforms accountable for algorithmic transparency and disinformation.
- **The Digital Markets Act (DMA):** challenging the monopolistic practices of Big Tech "gatekeepers."
- **The proposed AI Act:** classifying AI systems by risk level, banning harmful use cases entirely.

Together, these frameworks signaled that GDPR was not a one-off event. It was the foundation for something larger: a kind of digital constitution, one that framed data rights as civic rights and ethical AI as a public good.

Why This Case Study Matters
The EU's journey with GDPR offers a template for rights-based digital governance in a world increasingly defined by technological disruption. It underscores a central diplomatic lesson: that in the absence of global norms, regional leadership can set the tone.

Europe demonstrated that even in a domain dominated by tech titans, law can be a tool of values. That a regulation written in Brussels could ripple across Silicon Valley. That democracy and digital sovereignty don't have to be opposites—they can be intertwined.

Conclusion: Trust as Infrastructure
The GDPR's legacy is about more than consent boxes and privacy policies. It's about what kind of internet we choose to build—and who it serves. In the EU's view, digital infrastructure is not only about speed or bandwidth. It's also about trust. And trust, once institutionalized, can be just as powerful as code.

In the age of machines, the European Union offered something rare: a moral compass.

That compass may not point to a perfect destination. But in a world hurtling forward without a map, it offers direction—and that may be the most important kind of leadership there is.

CASE STUDY 2: CHINA AND THE DOCTRINE OF CYBER SOVEREIGNTY

Power, Control, and the Global Export of a Digital Worldview

In the global conversation on digital rights, many voices call for freedom—freedom of speech, movement, privacy, and innovation. But China's voice has long sung a different tune. Where others see the internet as a decentralized, open commons, China sees it as territory to be governed—bordered, surveilled, and controlled.

This philosophy has a name: cyber sovereignty. For the People's Republic of China, digital space is not a global playground but a sovereign extension of the state—no different than airspace or national borders. And over the last two decades, China has meticulously built a system that makes this philosophy not just theory, but reality.

It's a model built on control—and it works.

A System of Digital Walls and Eyes

China's internal architecture of cyber sovereignty is vast, complex, and ruthlessly efficient.

- The **Great Firewall** censors access to information, blocks foreign platforms like Google and Facebook, and filters search results based on government-approved narratives.
- All data generated in China must be stored **domestically**, on servers physically located inside the country.
- Both **foreign and local companies** are legally obligated to provide the state access to data deemed vital to national interests.
- A growing **social credit system** rates individuals on their online and offline behavior—rewarding those who comply, punishing those who dissent.

Every click, transaction, or post can be monitored. But for the Chinese government, this is not repression—it's governance. The system is framed as a mechanism for national stability, digital safety, and cultural sovereignty in a world seen as chaotic and Western-dominated.

From Domestic Model to Global Blueprint
What makes China's case extraordinary is not just how it governs its own cyberspace—but how it exports that model abroad.

Through the **Digital Silk Road**, a branch of its Belt and Road Initiative, China has:

- **Built digital infrastructure**—from fiber optic cables to surveillance networks—in dozens of countries.
- **Trained foreign officials** in cybersecurity, state media control, and algorithmic surveillance.
- **Sold facial recognition systems**, AI-enabled "smart city" packages, and cloud data centers that enable total visibility over populations.

From Southeast Asia to Sub-Saharan Africa, authoritarian-leaning governments have found a willing partner in China—not just in technology, but in ideology.

It's a powerful pitch: **stability over chaos, efficiency over debate, sovereignty over interference.**

A Digital Cold War of Values

The rise of China's cyber sovereignty model presents a clear and escalating contrast with Western ideals. On one side stands the open internet: flawed, unruly, and vulnerable to misinformation—but rooted in liberal principles. On the other, a tightly controlled ecosystem where expression is managed, surveillance is normalized, and dissent is algorithmically suppressed.

This isn't just about technology—it's about global governance. It's about what kind of world we want to live in.

Countries now face an implicit choice between competing digital futures:

- Do they adopt the EU's human rights–driven approach?
- Do they embrace Silicon Valley's innovation-first libertarianism?
- Or do they choose Beijing's model of digitally empowered authoritarianism?

This is the geopolitical fault line of the digital age.

Why This Case Study Matters

China's model of cyber sovereignty matters because it is working—both domestically and internationally.

- Domestically, it allows the state to preempt unrest, control narratives, and maintain loyalty.
- Internationally, it's being adopted in varying degrees by governments hungry for technological modernization without democratic reform.

This poses a profound challenge:

- Will the global internet fragment into "digital blocs" based on governance style?
- Can liberal democracies compete with the Chinese model without compromising their own values?
- Are human rights negotiable in exchange for security and efficiency?

As more nations face these questions, China's model becomes not just an alternative—but a serious contender for the future foundation of global digital governance.

Conclusion: The Code Behind the Curtain

China's doctrine of cyber sovereignty reminds us that the internet is not neutral. It is a mirror of the powers that govern it. Behind every firewall, facial recognition system, and biometric scan is a political decision—a view of the human being and the role of the state.

In China, the state is the architect. The user is a subject, not a citizen.

This may not be the digital future most of the world envisioned in the 1990s, but it is the one gaining momentum.

Understanding China's cyber governance isn't just about understanding a single country. It's about recognizing that digital models carry moral cargo. And in this new era of geopolitical contestation, **the battle for the soul of the internet is already underway.**

CASE STUDY 3: INDIA'S TIGHTROPE BETWEEN DEMOCRACY AND DATA NATIONALISM

Balancing Sovereignty, Innovation, and Rights in the World's Largest Democracy

India—home to over 1.4 billion people and one of the most dynamic technology sectors on Earth—is a digital giant on the move. Every day, millions of Indians come online, open bank accounts via their phones, access government services remotely, and engage with a rapidly expanding digital economy.

And yet, behind this transformation lies a difficult question: How does the world's largest democracy protect its people's data, preserve its national interests, and uphold individual freedoms—without sliding into digital authoritarianism?

India's journey is one of bold ambition, occasional contradiction, and constant negotiation. It is a test case not only for itself, but for the Global South—a model that could be replicated across dozens of nations facing similar trade-offs between growth, control, and liberty.

The Digital Leap Forward

India's digital revolution began with infrastructure, but quickly evolved into identity.

- **Aadhaar**, the world's largest biometric ID system, assigned each resident a unique number linked to fingerprints and iris scans—integrated into everything from banking to welfare.
- The **Unified Payments Interface (UPI)** revolutionized cashless transactions, enabling even small vendors and farmers to engage in digital commerce.
- Public services—from subsidies to vaccination records—went digital, sometimes before citizens were ready.

To many, this shift was liberating. It promised inclusion, especially for rural and low-income populations historically excluded from formal systems. Millions opened bank accounts, received government support directly, and gained a digital voice.

But that voice soon found itself in tension with the growing reach of the state.

Data Nationalism Rises

As India's digital footprint expanded, so did concerns over sovereignty, security, and foreign influence—especially in the wake of global data scandals and rising geopolitical tensions.

The Indian government responded with a push toward data localization—requiring that all personal data about Indian citizens be stored within the country. Officials argued that this was essential for protecting national security and creating digital independence.

The sentiment culminated in dramatic moves, such as:

- **Banning over 200 Chinese apps**, including TikTok and WeChat, citing national security concerns.
- Promoting **Indian-built alternatives** to dominant U.S. and Chinese platforms.
- Advancing the **Personal Data Protection Bill**, which, while meant to safeguard data, also included clauses that allowed government access to private information without a warrant.

These efforts, often labeled as "digital sovereignty," resonated with many Indians. But they also raised red flags.

Was the state protecting the people—or watching them?

Between Protection and Overreach

In 2017, India's Supreme Court declared that privacy is a fundamental constitutional right. This landmark judgment came in response to growing public unease about Aadhaar, which had become mandatory for everything from school enrollment to SIM cards.

Civil society groups, journalists, and technologists launched campaigns against what they saw as a creeping surveillance state. Their concerns were clear:

- **What safeguards exist** to prevent misuse of biometric and financial data?
- **Who holds the state accountable** when it collects data on citizens?
- **Is consent meaningful** if opting out means losing access to basic services?

Public debate intensified. Town halls, petitions, open letters, and protests became fixtures in India's digital discourse. The result wasn't always policy change—but it was proof that the democratic impulse was alive and pushing back.

Why This Case Matters

India's digital journey is unique in scale, but not in kind. Its experience echoes across the Global South: the allure of digital tools, the urgency of national control, and the fragility of rights in the face of rapid change.

What makes India's case critical is its symbolism:

- It is a **democracy**, not an authoritarian state.
- It is a **technology leader**, not merely a consumer.
- It is a **global voice**, not a passive participant in digital governance.

As such, the way India balances its national aspirations with the rights of its people will shape the choices of other nations struggling to walk the same tightrope.

Conclusion: Rights at the Speed of Reform

India is writing a digital script for the 21st century— one line at a time, with all the noise, debate, and contradiction that democracy entails. It may not always move fast, but it moves with fierce public engagement.

Whether it ultimately becomes a beacon for digital democracy or slides toward digital control depends not just on laws—but on culture, transparency, and trust.

India reminds us that **data is not just about economics or security—it is about citizenship**. And in the digital age, the right to be protected must include the right to not be watched.

This is why India's path matters. Not just for itself, but for a world learning that **sovereignty must be earned through consent—not claimed by code.**

DIPLOMACY IN A WORLD WITHOUT BORDERS

As data becomes a source of national strength, diplomacy must adapt. Negotiating over data transfers, digital trade, cybersecurity cooperation, and algorithmic accountability has become as important as negotiating over arms or energy.

Yet the digital world resists simple rules. Data crosses borders in milliseconds. Platforms operate in dozens of jurisdictions. Algorithms evolve faster than treaties can be drafted. And the traditional tools of diplomacy—bilateral meetings, communiqués, summits—often feel ill-suited for such a fluid terrain.

Still, diplomats are rising to the challenge. New coalitions are forming to craft norms for responsible behavior in cyberspace. Countries are engaging with tech firms as quasi-sovereign actors, seeking agreements not just between governments, but between platforms and the public interest.

This is a new kind of diplomacy—technical, ethical, fast-moving, and deeply human. It requires not only legal expertise, but cultural fluency, digital literacy, and a willingness to confront uncomfortable truths about how power works in a connected world.

CONCLUSION: SOVEREIGNTY IN THE AGE OF DIGITAL IDENTITY

The battle over data is not merely technical—it is profoundly human. It is a struggle over how we are seen, remembered, governed, and ultimately valued. It touches the core of who we are, how we relate to institutions, and how we claim agency in a digital world that is constantly watching, sorting, and monetizing our existence.

When the European Union introduced GDPR, it did more than regulate. It sent a message to the world that personal data is an extension of personal dignity. That consent is not optional. That individuals are not products, and privacy is not a privilege—it is a right. That was Europe's way of drawing a line in the digital sand.

In contrast, China's model of cyber sovereignty reframes the internet not as a space for freedom, but as an extension of national power and ideological control. It promises safety through surveillance, harmony through restriction, and order through censorship. It is efficient, it is strategic, and it is spreading—offered as a digital governance solution to governments grappling with instability and dissent.

India, standing between these poles, reflects the tension faced by every democracy in the Global South: how to maintain control over its data and platforms without betraying its pluralism and freedoms. Its path is still uncertain—but its choices will have enormous influence on how the next billion people experience rights, identity, and power in the digital age.

All three models raise critical questions:

- Can sovereignty protect the person as well as the state?
- Can regulation be both just and innovative?
- Can democracies remain open while defending against weaponized disinformation and foreign interference?

This is not just a policy debate. It is a test of vision and will.

In the coming years, the nations that lead in digital diplomacy and governance will not be those with the most advanced algorithms or surveillance tools—but those with the courage to place human dignity at the center of their digital architecture.

Because in the end, sovereignty is not only about borders or bandwidth. It is about who controls the narrative of our lives—who decides what is remembered, what is erased, and what is monetized. In a world where data is the new currency and narrative is the new power, defending digital autonomy is an act of both security and solidarity.

We must build a digital future where memory is not exploited, identity is not weaponized, and sovereignty serves the people—not just the powerful.

And in that future, the most sacred ground a nation can defend may no longer be land—but the digital soul of its people.

Chapter 11: The Evolving Role of Non-State Actors in Digital Diplomacy

WHEN THE WORLD SPEAKS BEYOND THE STATE, DIPLOMACY MUST LEARN TO LISTEN

The architecture of diplomacy is undergoing a quiet revolution.

For centuries, international affairs were the domain of states. Sovereign governments negotiated treaties, managed crises, and shaped global norms. Embassies, flags, and official titles defined who had the right to speak on behalf of a people or a nation.

But the digital age has unraveled that exclusivity.

Today, the power to shape narratives, influence public opinion, and even alter diplomatic outcomes is no longer reserved for ministries of foreign affairs. It belongs also to tech giants, youth activists, online communities, nonprofit organizations, investigative journalists—and even anonymous users with a smartphone and a signal.

We have entered a new era: one of distributed diplomacy, where state and non-state actors share the same global stage.

This chapter explores the implications of that shift—how diplomacy is being transformed by non-state actors, why their influence matters, and how states must adapt to remain relevant in a conversation they no longer fully control.

DIPLOMACY BEYOND BORDERS AND BADGES

The notion of diplomacy once meant closed-door negotiations, staged summits, and deliberate protocol. Now, diplomacy unfolds in **real time**, often in front of millions.

- A CEO's tweet can trigger a currency dip.
- An activist's Instagram post can shift international funding.

- A fact-checked video from a nonprofit can challenge an official government statement.

These aren't fringe outliers—they are evidence of a deeper trend: **the decentralization of influence**.

In the past, legitimacy in diplomacy came from position. Today, it comes increasingly from **impact, access, and authenticity**.

WHO ARE THE NON-STATE ACTORS OF THE DIGITAL AGE?

1. Technology Companies

Companies like Meta, Google, and X (formerly Twitter) are not just platforms—they are **infrastructures of diplomacy**. Their algorithms can elevate or suppress political speech. Their decisions can empower dissidents or embolden regimes.

These corporations now:

- Negotiate with governments
- Enforce content moderation policies with geopolitical impact
- Shape how billions access information

They have become unofficial ambassadors of the digital realm, often without a democratic mandate.

2. Civil Society and Activist Movements

From the Arab Spring to Black Lives Matter to #MeToo, social movements fueled by digital tools have shaped global awareness, spurred diplomatic action, and even influenced foreign aid policies.

Their power lies in storytelling—authentic, emotional, urgent—and in their ability to build transnational coalitions that pressure states to act or reform.

3. NGOs and Journalistic Entities

Think tanks, international watchdogs, and independent media organizations now conduct investigations, publish reports, and convene diplomatic dialogues that influence state behavior.

These actors serve as truth-tellers and conveners, often trusted more than governments themselves.

4. Hackers, Leakers, and Digital Whistleblowers

Not all non-state actors operate in sunlight. Some wield influence through exposure—releasing classified information, uncovering corruption, or disrupting surveillance regimes.

While controversial, they've forced governments to confront uncomfortable truths and raised public awareness around privacy, surveillance, and democratic accountability.

5. Influencers and Cultural Diplomats

In the age of followers and feeds, cultural icons and digital creators can sway opinion more effectively than official statements.

Whether advocating for humanitarian causes or shaping perceptions of conflict, these individuals act as unofficial voices of diplomacy, humanizing abstract crises.

EXPANDED CASE STUDIES: THE IMPACT OF NON-STATE DIGITAL DIPLOMATS

Case Study 1: Meta and the Rohingya Crisis

When Silence Becomes Complicity

In the rural towns and bustling cities of Myanmar, Facebook wasn't just a platform—it was the internet.

For millions, especially in areas with limited media literacy and few alternative news sources, the blue icon on their phones became the lens through which they saw the world.

In 2017, that lens turned violent.

The Myanmar military, already deeply embedded in the country's fragile political transition, recognized the power of Facebook as a megaphone for influence. Through coordinated campaigns, fake accounts, and nationalist rhetoric, the platform became flooded with anti-Rohingya propaganda. Posts described the Muslim minority as "invaders," "terrorists," and "threats to Buddhism." Graphic images—often doctored—spread at viral speed. Calls to violence were not subtle. They were explicit.

What followed was one of the gravest humanitarian tragedies of the digital era. Entire Rohingya villages were burned to the ground. Survivors spoke of mass killings, rape, and forced displacement. By the end of 2017, more than 700,000 Rohingya had fled across the border to Bangladesh. The United Nations called it ethnic cleansing.

Behind the human suffering was a chilling digital trail. And at the center of that trail stood Facebook—now Meta—a company that had built an unprecedented communication network but failed to anticipate or stop its weaponization.

At the time, Meta had no local language content moderators in Myanmar. Its algorithms, tuned for engagement, promoted the most inflammatory content. Civil society warnings went unanswered. The company, in pursuit of growth, had ignored the warning signs. Only after international investigations and media pressure did Meta admit its failure.

In a 2018 report commissioned by Meta itself, the conclusion was damning: Facebook had been used to "foment division and incite offline violence." The platform, by failing to act, had become complicit—not in intent, but in consequence.

Diplomatic ripples followed swiftly. Human rights groups filed lawsuits. The UN's Independent International Fact-Finding Mission on Myanmar named Facebook as having played a "determining role" in the crisis. Pressure mounted not only on the company, but on governments to regulate how social media platforms operate in volatile environments.

Meta responded with reforms. It hired more Burmese-speaking moderators, partnered with local NGOs, and rolled out AI tools to detect hate speech. But for many, the damage was done. Trust was shattered. A platform once marketed as a bridge between people had, in Myanmar, become a trigger for genocide.

This case became a turning point in the conversation about platform responsibility. It raised urgent questions: Can a tech company be held accountable for atrocities it failed to prevent? Should social media platforms be treated as neutral infrastructure—or as diplomatic actors with global influence?

In the aftermath, Meta acknowledged its role. But acknowledgment was only the beginning. The Rohingya crisis remains a haunting example of what happens when global technology collides with fragile societies—and no one is watching.

Case Study 2: Greta Thunberg

The Climate Diplomat with No Passport Stamp

It began with one girl. One cardboard sign. One quiet protest outside the Swedish Parliament.

Greta Thunberg was just fifteen years old when she skipped school on a crisp August morning in 2018 to sit, alone, with a hand-painted sign that read: Skolstrejk för klimatet—School Strike for Climate. She didn't shout. She didn't storm buildings or hold a microphone. She simply sat in silence, with fierce eyes and a sense of purpose that would soon ignite a global movement.

By the end of that year, thousands of students across Europe were striking on Fridays. By 2019, the movement—Fridays for Future—had swept across more than 150 countries, drawing millions into the streets. What began as a solitary act of defiance had transformed into one of the most urgent youth-led political mobilizations of the 21st century.

And at the center of it all was Greta.

She was not a diplomat in any formal sense. She carried no official portfolio. She had no staff, no embassy, no government backing her. But Greta wielded a different kind of power—moral clarity. Her voice, trembling yet resolute, pierced through the cluttered language of global diplomacy. She spoke in facts and feeling, in science and shame.

Her speeches became lightning rods.

At the UN Climate Action Summit in 2019, she looked world leaders in the eye and thundered: "How dare you?" Her voice cracked with fury and fear. "You have stolen my dreams and my childhood with your empty words." The room froze. The video went viral.

World leaders were unprepared—not for the science, but for the rawness. Greta refused the polite applause of diplomatic circles. She called climate summits a "celebration of blah blah blah," exposing the chasm between promises and action. She embarrassed governments not with data alone, but with the mirror she held up to their delay and denial.

But Greta didn't stop at speeches.

She sailed across the Atlantic in a zero-emission boat to avoid the carbon footprint of air travel. She refused awards when she believed they were distractions from action. She redirected media attention to underrepresented voices from the Global South, reminding the world that climate justice was also racial, economic, and generational justice.

Diplomats, negotiators, and UN officials began to take notice. Youth delegations were no longer a token presence—they became essential. Climate negotiations began to shift in tone, forced to address not only metrics and timelines, but the ethical burden of inaction. Words like "intergenerational equity" entered official communiqués. The urgency of the youth voice could no longer be sidelined.

Greta had no flag, but she carried a cause. She had no passport stamps to show for statecraft, but her activism crossed borders with more impact than many official envoys. She spoke to power not from within it, but from outside—exposing its failures and reminding it of its responsibilities.

And through it all, she never claimed to be a leader. "I'm just a messenger," she often said. "The science is the message."

Yet in the halls of power, in the streets of capital cities, and across countless social media feeds, Greta's message became impossible to ignore. She forced climate diplomacy to confront its own conscience. And she gave voice to a generation that had inherited a problem it did not create—but refused to accept.

Case Study 3: Bellingcat

Citizen Intelligence Changing the Global Narrative

They don't wear uniforms. They don't carry diplomatic titles. Most of them work from kitchen tables, spare bedrooms, or shared offices scattered across the globe. But in today's information wars, few forces have been as disruptive—or as quietly revolutionary—as Bellingcat.

Founded in 2014 by British blogger Eliot Higgins, Bellingcat began as a one-man experiment. Higgins, unemployed and armed with nothing more than a laptop, started analyzing videos from the Syrian civil war. He noticed patterns—missile fragments, terrain features, timestamps. While media outlets and governments struggled to verify footage, Higgins was geolocating airstrikes and matching artillery types to grainy YouTube clips.

Soon, others joined him. Not journalists in the traditional sense, nor intelligence officers, but digital detectives—people with a sharp eye, a strong sense of justice, and a belief that truth could be uncovered in pixels and metadata. They called themselves open-source investigators. The world began calling them something else: game-changers.

Bellingcat's turning point came with the downing of Malaysia Airlines Flight MH17 over eastern Ukraine in July 2014. While Russia and its proxies peddled alternative theories—blaming Ukraine, NATO, even suggesting a mid-air explosion—Bellingcat quietly collected satellite imagery, social media posts, and photos uploaded by locals. Their team traced the missile launcher, a Russian Buk system, from a base inside Russia to the Ukrainian border and back again. The evidence was overwhelming. Their findings predated official investigations—and held up in court.

From that moment, Bellingcat was no longer just a fringe group of hobbyists. They had stepped into the world of geopolitical consequence.

Their work continued to pierce the veils of state secrecy. In 2018, when Russian agents poisoned former spy Sergei Skripal in Salisbury, UK, Bellingcat went digging. Using leaked Russian passport databases, hotel records, and photos cross-referenced with public data, they identified the GRU agents involved—before any government did. Then came the poisoning of Alexei Navalny. Once again, Bellingcat led the way, exposing the Russian FSB team that shadowed Navalny across the country, piecing together their routes with the precision of a state agency—only without the resources, the mandate, or the official protection.

What set Bellingcat apart wasn't just what they uncovered—it was how they did it.

Everything was transparent. Every source was public. Every method was replicable. Their investigations could be checked, challenged, and verified. In an age of deepfakes, fake news, and digital cynicism, Bellingcat offered something rare: radical transparency.

- Their model changed the rules.

Where once intelligence was the domain of spies and governments, now a teenager with a laptop and a curious mind could expose state secrets. Bellingcat trained citizen journalists, ran workshops, and empowered grassroots watchdogs around the world. In Syria, Colombia, Sudan, and Yemen, open-source investigators used similar techniques to document war crimes and human rights abuses.

But their rise also sparked backlash.

Authoritarian governments branded them as tools of Western influence. Russian media labeled them provocateurs. Bellingcat's members faced hacking attempts, harassment, and smear campaigns. Yet they persisted—because the stakes were higher than reputation.

In courtrooms across Europe, their evidence informed war crimes charges. At the UN, their findings were cited in human rights reports. In living rooms and classrooms, their work inspired a new generation of investigative citizens who believed that facts, when made visible, could still change minds—and history.

The implications for diplomacy are profound.

Bellingcat did not wait for permission. They didn't issue press releases or seek press credentials. They simply uncovered truth and let it speak. For governments, this meant no longer controlling the narrative through traditional channels. For foreign ministries, it required grappling with uncomfortable facts, revealed not by whistleblowers or defectors—but by ordinary people using public tools.

In a world where secrecy used to be power, Bellingcat proved that openness could be just as potent.

Their legacy is still unfolding. But one thing is certain: the next diplomatic crisis may not be broken by a cable or a press conference—but by a tweet, a TikTok video, or a Bellingcat thread unraveling a deception in real time.

CHALLENGES AND ETHICAL TENSIONS

1. Legitimacy Without Accountability

Non-state actors can influence global policy without democratic oversight. Their platforms are massive—but who governs their conduct?

2. Information Saturation and Truth Fatigue

Multiple narratives competing for attention can lead to confusion, polarization, and distrust in all sources—state and non-state alike.

3. Manipulation Risks

Bad-faith actors—troll farms, bots, fake NGOs—can exploit the same digital tools for disinformation, eroding public trust and distorting diplomacy.

4. Inequitable Access

Not all voices have equal reach. Global South actors, Indigenous leaders, and rural communities may still be underrepresented in digital diplomacy—a gap that must be closed.

TOWARD COLLABORATIVE DIPLOMACY

The rise of non-state actors doesn't signal the end of traditional diplomacy—it signals the need for evolution.

Diplomats must:
- **Engage with civil society**, not just tolerate it.
- **Partner with platforms** to co-design digital peacebuilding and content governance.

- **Listen to digital movements**, not only formal delegations.

- **Incorporate non-state actors** into norm-building, humanitarian response, and technology ethics.

The diplomacy of the future is not exclusive. It is **inclusive, networked, and participatory.**

CONCLUSION: A SHARED TABLE, A SHARED RESPONSIBILITY

Diplomacy, once confined to embassies, backchannel telegrams, and elite gatherings, has been cracked open. Its architecture is no longer built solely by states, but co-constructed—line by line, post by post—by citizens, corporations, platforms, and movements. In the digital age, power doesn't just sit in presidential palaces. It pulses through timelines, hashtags, open-source maps, and viral speeches from voices who never needed a passport to be heard.

We are witnessing a redefinition of what it means to be a diplomat. It is no longer only the preserve of ambassadors in tailored suits or negotiators at long mahogany tables. Today, the activist live-streaming protests, the coder verifying war crimes, the influencer galvanizing climate action, and the platform engineer deciding what to remove and what to amplify—all are playing roles in global affairs. Some are aware of their influence. Others wield it unknowingly.

This is a moment of great opportunity—and great risk.

When states ignore non-state voices, they risk irrelevance. When platforms fail to recognize their geopolitical weight, they become conduits for harm. When movements rise without structure or safeguards, they risk being co-opted or silenced. The future of diplomacy depends on navigating these tensions with humility, clarity, and courage.

And yet, the promise is profound.

Imagine a diplomacy where governments co-create solutions with youth movements, where tech companies design systems with ethical foresight, where civil society is not an afterthought but a strategic partner. A world where truth is defended collaboratively, and where global challenges—climate change, pandemics, disinformation, inequality—are met not by a few actors behind closed doors, but by an open, collective architecture of accountability and care.

In such a world, diplomacy becomes something deeper than negotiation.

It becomes shared stewardship of our common future.

The table is already being reshaped. Its legs rest not only on sovereignty and strategy, but on transparency, empathy, inclusion, and participation. The question is no longer whether governments will share it—but how willingly, and with whom.

And for each of us, another question remains:

In a world where every post, every protest, every platform decision shapes diplomacy— **What kind of diplomat will you choose to be?**

Part III: Identity, Rights & Inclusion

Chapter 12: Gender and Digital Diplomacy

"Digital diplomacy is not gender-neutral. It either reinforces exclusion – or reimagines inclusion."

INTRODUCTION: DIPLOMACY HAS NEVER BEEN NEUTRAL

For centuries, diplomacy was built in the image of patriarchal power: exclusive, hierarchical, and overwhelmingly male. Women entered late, often as interpreters of emotion or symbols of cultural grace, but rarely as architects of foreign policy. The digital transformation of diplomacy has created new spaces for participation—but also new forms of vulnerability and violence.

"Digital diplomacy is not gender-neutral," notes Dr. Sanjana Verma, a gender policy expert at the UN. "It either reinforces the existing imbalance—or it dares to redesign the system." This chapter explores how gender and digital diplomacy intersect in real and consequential ways—from who gets heard to who gets harassed, from whose stories go viral to whose safety is compromised.

In the 21st century, gender is not just a lens—it is a fault line. And diplomacy, if it is to remain credible, must reckon with it.

THE DIGITAL DIVIDE IS GENDERED

Opportunities:

- Women and gender-diverse leaders are using digital platforms to amplify marginalized voices and bypass institutional gatekeepers.
- Movements like #MeToo, #BringBackOurGirls, and #WomanLifeFreedom have shown the ability of digital activism to influence policy.
- Digital diplomacy creates new routes to visibility, especially in societies where women have limited offline representation.

Challenges:

- Online gender-based violence disproportionately affects women in diplomacy and advocacy. A UN Women report (2021) found that 73% of female politicians globally had experienced cyber harassment.

- Structural underrepresentation in cybersecurity, AI ethics, and tech policy spaces means that women often have limited influence over the rules that govern digital diplomacy.
- Intersectionality is often ignored. Indigenous, queer, disabled, and non-Western women are excluded both from diplomatic stages and digital advocacy campaigns.
- As Nanjira Sambuli, a Kenyan digital rights advocate, puts it: "If we replicate offline exclusion online, we are not building a new world—we are just digitizing injustice."

EXPANDED CASE STUDIES: GENDER AND DIGITAL INFLUENCE IN ACTION

Case Study 1: Sweden's Feminist Foreign Policy and Digital Messaging Strategy

When Sweden announced its feminist foreign policy in 2014, it did more than issue a political statement—it launched a quiet revolution in how diplomacy could be practiced and communicated in the digital age. At a time when gender was often treated as a secondary concern in international affairs, Sweden positioned it at the core of its foreign engagements. The country's ambition was not only to advocate for women's rights globally but to embody feminist principles through all its diplomatic institutions and actions—including how it communicated online.

Margot Wallström, the then-Foreign Minister, described the initiative as "a foreign policy that fights for women's rights, representation, and resources—globally and unapologetically." But how would that translate in practice, especially on platforms historically shaped by male-dominated discourse and algorithmic bias?

Sweden's Ministry for Foreign Affairs began experimenting with digital storytelling as a tool of soft power. The flagship initiative, "She Leads," was more than a hashtag. It was a coordinated campaign across embassies, ministries, and consulates, spotlighting women leaders from all walks of life: a climate activist in the Pacific Islands, a judge in Rwanda, a coder in Egypt. These profiles were shared with short, compelling captions, often backed by data: for example, noting how conflict zones with greater gender parity in peace processes saw more lasting peace agreements.

"We decided that diplomacy should not just speak—it should listen, reflect, and uplift voices rarely heard," said Annika Söder, then State Secretary for Foreign Affairs. "Digital platforms gave us the ability to be more than formal. We could be human."

The initiative's digital reach was significant:
- **Over 10 million engagements** in its first year;
- **Embassy-led campaigns** in Kenya, Mexico, and Indonesia encouraging local participation;
- Integration of **gender narratives** in climate, trade, and peacebuilding dialogues.

Sweden's bold stance didn't go unnoticed. Within a few years, countries such as **Canada, France, and Mexico** followed suit, launching their own feminist foreign policy frameworks. Sweden had not only led with words but with design—and other states took note.

Still, the approach faced its share of backlash. Conservative governments viewed Sweden's messaging as ideological overreach. Critics questioned whether feminist branding truly translated into structural change, especially when Sweden's arms trade or domestic immigration policies came under scrutiny.

Inside the Ministry, this criticism led to introspection—and reform:

- **Training programs** were launched for diplomats on gender-sensitive communication;
- **Internal reviews** ensured consistency between external messaging and policy practice;
- Social media teams were equipped **with tools to manage backlash** and promote inclusive dialogue.

Despite challenges, Sweden's case offers a powerful lesson: **digital platforms can enhance feminist diplomacy, but only when they are backed by credibility, transparency, and consistency.** When used responsibly, these platforms allow diplomacy to move from closed rooms to open networks, from statements to stories, and from policy to presence.

Case Study 2: Harassed Into Silence

Women Diplomats in the Global South

Amina, a 29-year-old junior diplomat from East Africa, was no stranger to digital engagement. She had helped design her country's virtual strategy for engaging in UN peacebuilding forums. But in early 2020, after appearing on a livestreamed panel hosted by the UN on women in post-conflict negotiations, her life changed overnight.

Shortly after the video was posted online, Amina began receiving a flood of messages. At first, they were critical. Then they turned hostile. By the end of the week, she had received **rape threats, doctored photos, and personal doxxing**. Her professional accounts were overwhelmed with disinformation suggesting she was a traitor. Her private phone number had been leaked.

"I knew harassment was a risk," she said later. "But I never imagined the scale, or how fast it would escalate."

Amina's experience is not isolated. Across the Global South, women diplomats, journalists, and civil society leaders face **coordinated online attacks**—particularly when they speak about topics seen as controversial: LGBTQ+ rights, post-colonial critique, gender-based violence, or inclusion in peace processes.

Key patterns of abuse:
- **Organized campaigns** often led by bot networks or far-right groups;

- **Targeted deepfakes** and misinformation to discredit professional credentials;
- **Platform inaction**, with slow or no removal of violent threats.

In Colombia, trans diplomat Camila Andrade faced similar abuse after participating in an OAS conference on gender-inclusive conflict resolution. "They attacked not just my views," she said, "but my identity. My very right to exist."

In South Asia, human rights defender Sana Malik was targeted with deepfakes after a UN climate conference where she advocated for gender parity in environmental negotiations.

Reports from organizations like **Amnesty International** and **Access Now** confirm a disturbing trend: gendered disinformation and online violence are being weaponized to silence women, particularly from the Global South. The attacks are often organized by **bot networks, far-right groups**, or even state-aligned actors. Platforms such as X (formerly Twitter), Facebook, and YouTube have been slow to respond.

Consequences for diplomacy:

- Women retreat from public discourse due to fear;
- Career advancement is hindered by exposure to harm;
- Institutions lack the infrastructure to provide support.

However, grassroots organizations have stepped in. The **Digital Rights Foundation** in Pakistan, **Luchadoras** in Mexico, and **Pollicy** in Uganda have all created resources—from cyber hygiene workshops to crisis response teams. The **Digital Defenders Partnership** now funds emergency assistance for women at risk, offering secure devices, legal support, and even relocation if needed.

Some governments are catching up:

- **Kenya** launched a national task force on digital harms;
- **Argentina** is developing public protocols with feminist tech groups.

This case reveals a hard truth: **digital diplomacy cannot be inclusive if it is not safe**. Security is not just about embassies and encryption—it is about the emotional and political resilience of those who speak truth to power online.

Case Study 3: Iran and the Digital Uprising of "Woman, Life, Freedom"

In September 2022, the death of Mahsa Amini, a 22-year-old Kurdish woman arrested by Iran's morality police, triggered a national and global outcry. What began as grief erupted into one of the most visible digital feminist uprisings of the modern era.

Videos flooded the internet: women cutting their hair in defiance of the regime, chanting in the streets, dancing without hijabs. Hashtags like **#MahsaAmini** and **#ZanZendegiAzadi (Woman, Life, Freedom)** dominated global social media for weeks.

Despite internet shutdowns and censorship, the movement quickly moved across borders:
- Diaspora communities organized protests in Berlin, Toronto, and Los Angeles;
- **Clubhouse rooms** and Twitter spaces brought together Iranian voices and international media;
- Online petitions and campaigns pushed **UN bodies and foreign ministries** to act.

Feminist diplomats responded:

- Canada, the EU, and the U.S. **imposed sanctions** on Iran's morality police;
- **Gender apartheid** became a term recognized in multilateral discussions;
- Activists like **Masih Alinejad** became digital envoys of the movement.

"This is not a social media moment," said Alinejad in a viral video. "This is a demand for justice, and it deserves consequences."

Tech platforms were also challenged. Meta, under pressure, reversed some content moderation policies and pledged greater transparency around Iran-related takedowns.

What made this protest different?
- It was leaderless yet coordinated;
- It combined grassroots courage with diplomatic traction;
- It showed that feminist resistance can shape foreign policy.

Though the Iranian regime responded with violence and repression, the protest movement fundamentally altered how digital feminist activism is seen on the world stage.

It proved that the center of diplomacy is shifting—from ministries to movements, from podiums to platforms. And that when justice finds a voice online, diplomacy must learn to listen—or risk becoming irrelevant.

TOWARD A GENDER-INCLUSIVE DIGITAL DIPLOMACY: A COMPACT AGENDA

To make diplomacy truly inclusive in the digital era, we must move beyond statements of intent and build structures that reflect equity, safety, and shared leadership. This begins with five core shifts:

1. **Embed Gender from the Start.** Every digital policy—whether on cybersecurity, AI, or data governance—must begin with a gender lens. If gender is not built into the design, the outcome will reinforce exclusion.

2. **Prioritize Safety as Strategy.** Online harassment isn't a side issue—it's a threat to participation.

Diplomatic institutions must create safety protocols, early-warning systems, and support networks for those targeted online.

3. **Fund Women-Led Innovation.** Support must go beyond visibility. Governments and multilateral bodies should invest in women-led platforms, policy labs, and civic tech—especially in the Global South.

4. **Train Allies, Especially Men.** Gender inclusion requires more than good intentions. Male diplomats must be equipped to model inclusive behavior and respond to digital abuse with action, not silence.

5. **Connect Across Sectors.** The most durable change happens in coalitions. Diplomats should actively collaborate with feminist technologists, civil society leaders, and youth networks to co-create just and resilient digital systems.

Inclusion isn't just a value—it's infrastructure. And without it, digital diplomacy will remain unequal by design.

CONCLUSION: FROM REPRESENTATION TO REIMAGINATION

In the digital age, gender is not a peripheral issue—it is a **structural determinant of voice, access, and impact.**

Digital diplomacy that fails to integrate a gender lens is not only outdated—it is dangerous, reinforcing the very hierarchies it should help dismantle.

The future of diplomacy must be:

- **Feminist** in its understanding of power
- **Intersectional** in its inclusion of diverse identities
- **Digitally conscious** in its design of safe, equitable platforms for engagement

We cannot create a secure digital future without ensuring it is also **just, representative, and human-centered.**

And as more women—and gender-diverse leaders—step into digital diplomacy, they bring not only new voices but new visions for peace, justice, and planetary dignity.

Because diplomacy, at its core, is about building relationships.

And relationships that do not include **half the world's perspectives** will never be strong enough to carry the weight of tomorrow.

Chapter 13: The Role of Culture in Digital Diplomacy

CULTURE IS NOT AN ACCESSORY OF DIPLOMACY. IT IS ITS SOUL

In every era of global affairs, culture has been both a mirror and a message. Long before treaties were signed in official halls, human societies exchanged ideas, art, music, and customs across borders and empires. Cultural exchange has always predated formal diplomacy. But in the 21st century—and especially in the digital age—culture has moved from the periphery to the center of global influence.

This is not a coincidence. It is a reflection of what people trust, what they remember, and what shapes how they perceive one another. In an age marked by information overload, political fatigue, and polarization, culture offers a different kind of currency: **authenticity, emotion, and meaning.**

Today, the reach of cultural diplomacy is unparalleled. It flows across borders not by decree but by demand. A viral video, a streaming documentary, a traditional recipe shared on Instagram—these are now acts of diplomacy, shaping how people and states understand one another.

This chapter explores the dynamic role that culture plays in digital diplomacy. It analyzes how soft power is projected and contested through digital mediums, what risks and opportunities emerge from cultural globalization, and how practitioners can ethically and strategically engage with culture in a connected world.

FROM SILK ROADS TO STREAMING PLATFORMS: A BRIEF HISTORY OF CULTURAL DIPLOMACY

Cultural diplomacy has always existed—whether in the form of royal emissaries exchanging gifts, scholars translating ancient texts, or nations showcasing their artists and athletes at world fairs or the Olympics. These were expressions of identity and influence.

In the 20th century, cultural diplomacy became more institutionalized. During the Cold War, both the U.S.

and USSR heavily invested in cultural programs—jazz tours, film exports, ballet performances—as tools of ideological competition. Culture was not ornamental; it was geopolitical strategy.

Today, we are witnessing a new stage of this evolution: the rise of **digitally distributed cultural diplomacy**. This form is not centrally controlled. It is shaped by artists, influencers, grassroots movements, and audiences who share, remix, and reinterpret content across digital ecosystems.

CULTURE AS SOFT POWER IN THE DIGITAL AGE

1. Redefining Soft Power

Soft power, a term coined by Joseph Nye, refers to a country's ability to influence others through **attraction rather than coercion**. Culture, values, and foreign policy all contribute to this form of power.

In digital diplomacy, soft power plays out in:
- **Global cultural exports** (e.g., music, TV, cuisine)
- **Language proliferation and education programs**
- **Digital storytelling and nation branding**

- **Diaspora and identity narratives**

What's changed is the speed, scale, and sponsor of this influence. No longer solely state-led, digital culture spreads organically, often independent of government intent.

2. When Culture Leads Where Diplomacy Cannot
Culture can enter spaces that formal diplomacy cannot:
- Conflict zones: where political envoys are barred but artists are welcomed.
- Sanctioned states: where film or music can transcend embargoes and touch hearts.
- Post-colonial contexts: where official apologies fall short, but cultural empathy opens healing.

Cultural diplomacy, done well, humanizes the foreign, fosters understanding, and builds bridges that endure beyond policy shifts.

RISKS AND COMPLEXITIES IN THE DIGITAL CULTURAL ARENA

1. Cultural Appropriation and Misrepresentation
The global flow of culture can lead to unintentional offense or exploitation:

- Sacred symbols commodified in fashion
- Traditional dances used out of context
- Minority cultures erased or mimicked without representation

Digital platforms magnify these tensions, as content is often decontextualized and monetized without consent.

2. Algorithmic Bias and Inequality

Algorithms often favor dominant languages and high-production content. As a result:

- Indigenous and minority voices are underrepresented
- Local content creators struggle for visibility
- Cultural diversity online becomes performative, not authentic

3. Weaponization of Culture

Culture can be used for manipulation:

- State-backed media producing propaganda as entertainment
- Identity-based narratives stoking division or radicalization

- Historical revisionism through meme culture and influencers

Thus, cultural diplomacy must be aware of its impact, not just its intention.

EXPANDED CASE STUDIES: CULTURAL DIPLOMACY IN ACTION

Case Study 1: South Korea's Digital Cultural Renaissance – The K-Wave

When you think of cultural diplomacy in the 21st century, few examples stand out as powerfully as South Korea's global soft power phenomenon known as Hallyu, or the Korean Wave. What started as a regional trend in pop culture has evolved into a worldwide movement—and behind its success is a strategic blend of government policy, private sector innovation, and cultural authenticity.

In the late 1990s, after experiencing the Asian financial crisis, South Korea realized that economic recovery and global influence could come not only from heavy industries but from cultural exports. The Ministry of Culture, Sports, and Tourism began to invest in music, film, food, language, and fashion. At the same time, Korean entertainment companies like SM, YG, and JYP nurtured talent in K-pop, building multimedia empires around groups like Girls' Generation, EXO, and BTS. These artists weren't just musicians—they became cultural diplomats.

As digital platforms like YouTube, Twitter, and V Live expanded globally, South Korea harnessed them to distribute its content with remarkable agility. Korean dramas were suddenly just a click away in Cairo or São Paulo. Fans connected with idols in real time, forming virtual communities that transcended borders. Digital content became both the vehicle and the venue of Korea's global cultural outreach.

Key milestones in the K-Wave's digital diplomacy:
- BTS became the first K-pop group to speak at the **United Nations**, delivering messages of self-love and inclusion.
- The Netflix hit **"Squid Game"** sparked global discussions on inequality, capitalism, and Korean society.
- Enrollments in **Korean language programs** and tourism to South Korea surged across continents.

"We didn't just sell music. We shared emotion, values, and our way of life," said Bang Si-hyuk, founder of HYBE Corporation. "That's what made the connection real."

The outcome? South Korea's image transformed from a tech-centric manufacturing nation to a creative, innovative, and emotionally resonant cultural power. Trade, education, and tourism boomed. South Korea signed new bilateral agreements, expanded language education programs, and opened Korean cultural centers in dozens of countries.

Despite criticism that the K-Wave risked becoming overly commercialized, its core message remained strong: Korea's identity is global, inclusive, and dynamic. And that identity is shaped and shared not only by the government—but by artists, fans, and storytellers alike.

Case Study 2: #AfricaToTheWorld –Cultural Assertion through Decentralized Storytelling

In stark contrast to South Korea's state-led cultural strategy, Africa's digital cultural diplomacy emerged organically—from the grassroots, driven by creators, influencers, and communities that refused to be defined by the outside world.

For decades, Africa's global representation was shaped by Western media, often reduced to narratives of poverty, conflict, or charity. But in the 2010s, a digital cultural revolution began reshaping this perception. It wasn't led by ministries or embassies—it was led by artists with smartphones and hashtags.

Musical genres like **Afrobeats, Afrofusion**, and **Amapiano** exploded onto global stages. Nigerian superstars like Burna Boy and Wizkid packed stadiums from London to New York. **Nollywood**, already the second-largest film industry in the world, began premiering on **Netflix**, with fresh voices like Kunle Afolayan redefining African storytelling. Fashion influencers from Lagos to Accra went viral, blending traditional motifs with streetwear swagger.

The digital realm became a new kind of borderless embassy:

- Hashtags like **#AfricaToTheWorld, #BlackIsKing**, and **#AfricanExcellence** unified creators and fans across the diaspora.

- Online platforms such as **YouTube, Instagram**, and **Audiomack** allowed local content to bypass traditional gatekeepers.
- Diaspora communities amplified the movement, turning concerts and film screenings into cultural gatherings.

"We stopped waiting for permission to be seen," said Tiwa Savage, a global music icon. "We created our own stage and invited the world to watch."

Impact on diplomacy:
- Africa's cultural diplomacy became a force for rebranding the continent—from recipients of aid to exporters of creativity.
- Informal cultural envoys replaced traditional diplomats, yet wielded global influence.
- Youth-led digital culture fostered **continental pride, entrepreneurship, and soft power** that caught the attention of embassies and policymakers.

Governments are now catching up. The African Union has recognized the creative economy as a strategic sector. Countries like Nigeria and Ghana are investing in cultural festivals and film funds.

This case proves a simple truth: when the state lags, people lead. **Digital platforms democratize diplomacy** by giving storytellers the power to narrate their own identity, on their own terms.

Case Study 3: Ukraine's Digital Cultural Defense in Wartime

In February 2022, as Russian forces invaded Ukraine, the world witnessed not only a geopolitical war—but a cultural one. Faced with the threat of erasure, Ukraine turned to its artists, historians, musicians, and technologists to defend not just territory, but identity.

As missiles fell on cities, museums scrambled to digitize archives. Cultural workers partnered with international institutions to **preserve artworks**, manuscripts, and historical data online. The **Ministry of Culture and Information Policy** launched campaigns to raise awareness of Ukraine's heritage—and the risks it faced.

But digital cultural diplomacy wasn't only about preservation. It was also about **narrative resistance**.

- Ukrainian musicians performed in bomb shelters and streamed their music globally.
- Writers and poets shared real-time updates, blending art and activism.
- President **Volodymyr Zelenskyy**, himself a former performer, delivered historic addresses that fused emotion with message.

"This is not just a war over land. It's a war over who we are, and whether the world will remember," said Andrey Kurkov, a Ukrainian novelist.

Global audiences responded:
- Cultural organizations around the world held benefit concerts and exhibitions to raise funds.
- Social media trended with Ukrainian language, symbols, and solidarity.
- Ukrainian resilience became a cultural emblem of freedom.

Key diplomatic impacts:

- The EU and UNESCO mobilized aid to protect cultural sites.

- Ukraine's digital presence—videos, posts, artworks—garnered empathy and policy support.
- The country's cultural identity became a moral anchor in the geopolitical narrative.

This case shows that in the face of war, **culture is not collateral—it is a line of defense**. Through digital channels, Ukraine told its story, claimed its space, and rallied the world—not through propaganda, but through human truth.

In an age of hybrid warfare, **digital cultural diplomacy has become an essential weapon—not of aggression, but of connection**.

STRATEGIC RECOMMENDATIONS FOR CULTURAL DIPLOMACY IN THE DIGITAL AGE

To integrate culture more effectively into digital diplomacy, practitioners and policymakers should:

1. **Embed Cultural Advisors in Digital Teams.** Include anthropologists, historians, and creatives in policy and public diplomacy units to ensure depth and context.

2. **Protect Digital Cultural Heritage.** Invest in cyber-archiving, decentralized backups, and global access to vulnerable collections—especially in conflict-prone or climate-affected regions.

3. **Fund Local and Indigenous Creators.** Shift from "representing" others to empowering them to speak for themselves, with funding and visibility.

4. **Establish Ethical Guidelines.** Develop cross-sector protocols for respectful cultural exchange, digital repurposing, and algorithmic equity.

5. **Promote Reciprocal Exchange.** Avoid extractive cultural diplomacy by building mutual platforms of exchange, co-creation, and long-term collaboration—not one-time showcases.

CONCLUSION: CULTURE IS THE MEMORY OF THE WORLD—AND THE PATH TO ITS FUTURE

In the clamor of geopolitics, culture is often dismissed as soft, decorative, or nonessential. But history tells a different story. It is culture that remains when borders shift. It is music and art that console when treaties fail.

It is storytelling that bridges generations and geographies alike.

In the digital era, culture is not only preserved—it is **performed, shared, shaped, and sometimes stolen.** It is vital that those guiding diplomacy understand not only its strategic value, but its sacred one.

Because when a song can reach more people than a summit, when a meme can ignite more thought than a manifesto, and when a story can build more empathy than a speech **culture becomes diplomacy.** Not as theater. But as truth.

And in a world desperate for meaning, those who carry culture are not just creators.

They are the **custodians of humanity's deepest form of connection.**

Chapter 14: Digital Identity, Citizenship, and Inclusion

"To be counted is to exist. In a digital world, identity is not just a name – it is a gateway to rights, to services, to belonging."

INTRODUCTION: WHO ARE YOU IN A BORDERLESS WORLD?

Identity used to be simple. It was a name on a passport, a fingerprint in a national archive, a photo laminated on a school ID card. But as the world moves online—across borders, platforms, and systems—identity has become something far more complex, and far more consequential.

Today, to lack a recognized digital identity is to be excluded. Without one, a person may be unable to access healthcare, enroll in school, open a bank account, vote, or receive aid. In many ways, **digital identity has become a prerequisite for citizenship itself—not just legal belonging, but functional**

inclusion in the economic, social, and political life of one's country.

But who defines this identity? Who controls the systems that issue it? And what happens when these systems fail—or are used to profile, exclude, or surveil?

This chapter explores how digital identity is reshaping global diplomacy and governance. From Estonia's e-Residency to the Emirates ID, from India's Aadhaar to UNHCR's refugee biometrics, the stakes are high. At its best, digital identity promotes access and empowerment. At its worst, it risks becoming a tool of control.

DIGITAL IDENTITY: THE NEW INFRASTRUCTURE OF CITIZENSHIP

At a fundamental level, digital identity is the data-based representation of a person—authenticated through biometrics, device records, blockchain credentials, or national databases. But beyond the technology, digital ID systems serve as gateways to rights.

They determine who receives government services, who is visible to international institutions, and who can participate in an increasingly digitized global economy. In a world where information replaces paperwork, digital identity is **the new passport, the new social contract, the new signature of existence**.

But identity is more than function—it's about dignity. The danger of digitizing identity without accountability is that people become entries in a database, disconnected from the power to shape how they are seen.

So how are different countries navigating this promise and peril?

CASE STUDY 1: ESTONIA'S E-RESIDENCY— REIMAGINING THE STATE AS A PLATFORM

Estonia, the tiny Baltic nation with just over a million citizens, has become one of the most ambitious pioneers of digital identity. After regaining independence in the 1990s, Estonia decided to leapfrog legacy systems and rebuild its state from scratch—this time, in the cloud.

By 2002, it had launched its national digital ID program. Every citizen and resident was issued a secure chip-enabled card that could be used for voting, healthcare, taxes, contracts, and more. But Estonia didn't stop at its borders. In 2014, it introduced e-Residency, allowing anyone in the world to apply for a government-issued digital identity.

E-residents are not citizens, but they can access Estonian services and even establish companies remotely. Entrepreneurs from Brazil to Bangladesh now use their Estonian e-IDs to manage EU-based businesses online.

Key elements of Estonia's e-Residency success:

- Built on transparency and strong cyber infrastructure.
- Partnership with global fintech and legal service providers to support e-residents.
- A strong narrative of digital sovereignty and democratic innovation.

The idea is revolutionary: a borderless identity system anchored in state legitimacy but accessible globally. It envisions the state not as a territory, but as a service provider in the digital era.

Impact:

- Over 100,000 e-residents from more than 160 countries.
- Thousands of new companies established in Estonia, contributing to its GDP.
- Positioned Estonia as a digital diplomacy leader in international forums.

Challenges remain:

- Security and AML/KYC risks.
- Sovereignty debates in cross-border regulation.
- Accessibility for digitally marginalized populations.

Critics question the model's potential for misuse or tax avoidance. But Estonia remains transparent in its governance and steadfast in its mission: to reimagine citizenship for a world where location is no longer destiny.

CASE STUDY 2: INDIA'S AADHAAR—SCALE, CONTROVERSY, AND INCLUSION AT THE EDGE

India's Aadhaar program is the largest biometric identity system in the world. Launched in 2009, it assigns a 12-digit unique ID to each resident, linked to fingerprints and iris scans. The goal was inclusion: to ensure that every Indian, especially the poor and marginalized, had proof of identity to access services.

Over 1.3 billion people are now enrolled. Aadhaar is used for subsidies, pensions, bank accounts, and more. It has streamlined welfare delivery, reduced duplication, and created one of the most extensive state-led digital systems globally.

Strengths of Aadhaar's system:
- Reduced fraud in government schemes.
- Enabled financial inclusion and mobile-based transactions.
- Interoperable across sectors and services.

But Aadhaar's journey has not been without friction. Civil liberties groups have warned of surveillance risks, exclusion errors, and mission creep—where Aadhaar was used beyond its original mandate, including by private entities. In some tragic cases, people reportedly died after being denied food rations due to authentication failures.

The 2018 Supreme Court ruling:
- Upheld Aadhaar's constitutional validity.
- Banned its mandatory use for services like mobile SIMs or school admissions.
- Affirmed the right to privacy as a fundamental right.

The court case became a landmark in global discussions on digital rights and surveillance. Aadhaar's story is a microcosm of the global identity dilemma: how to balance innovation with oversight, efficiency with equity, and reach with rights.

Lessons:
- Scale can amplify both impact and risk.
- Technical systems must be matched with grievance redressal.
- Identity is not only a number—it is a matter of dignity.

CASE STUDY 3: EMIRATES ID FOUNDATIONS FOR A SMART NATION

In the United Arab Emirates, digital identity has been integrated into nearly every dimension of daily life. The Emirates ID, issued by the Federal Authority for Identity, Citizenship, Customs and Port Security, is mandatory for all citizens and residents.

This smart card contains embedded biometric and demographic data and is required for accessing services such as healthcare, banking, government portals, mobile SIM registration, and even housing contracts. It also enables seamless interaction with the UAE's broader "smart government" agenda, where digital services are prioritized for speed, security, and innovation.

Key features of Emirates ID:
- Serves as a central authentication tool.
- Integrated into facial recognition and border control systems.
- Enables paperless governance and e-signature capabilities.

But what sets the Emirates ID apart is its alignment with the UAE's long-term vision for digital transformation. From AI-powered border control to blockchain-based public services, the Emirates ID serves as the central identity backbone in a society rapidly transitioning toward full e-governance.

Strategic implications:

- Strengthens public trust through efficiency and innovation.
- Serves as a foundational element in UAE's global competitiveness strategy.
- Positions the UAE as a regional leader in digital governance.

Critics raise concerns about data privacy, centralized control, and the absence of EU-style rights frameworks. Yet the system reflects a high-trust relationship between citizens and state—enabled by investment, infrastructure, and cultural orientation toward innovation.

In the UAE, digital identity is more than a tool—it is a national infrastructure for modernization and global competitiveness.

CASE STUDY 4: UNHCR'S BIOMETRIC IDENTITY FOR REFUGEES

Recognition Without a Nation

What happens when you have no state? For millions of refugees and stateless individuals, the question is not whether they can prove their identity—but whether they can be recognized at all.

The United Nations High Commissioner for Refugees (UNHCR) has launched biometric registration systems in camps across Africa, the Middle East, and Asia, aiming to provide each individual with a secure, portable digital ID. This system uses fingerprints, iris scans, and photos to create records that follow people even when borders do not.

Benefits of biometric identity for refugees:
- Enables access to aid, food distribution, and medical care.
- Reduces fraud and duplication in humanitarian systems.
- Establishes a digital paper trail in stateless contexts.

In theory, this gives displaced persons a way to access aid, protect against identity theft, and begin rebuilding their lives. In practice, it also introduces ethical questions: who owns the data? Can individuals delete their profiles? Are their biometric details secure?

In 2019, a data-sharing deal between UNHCR and Bangladesh over Rohingya refugees sparked outrage when it was revealed that biometric data may have been passed to the Myanmar government—without informed consent. For a community fleeing genocide, data sovereignty was not just a legal issue—it was a matter of life and death.

Critical reflections:

- Data protection is a human rights issue, especially for the vulnerable.
- Consent must be informed, transparent, and revocable.
- Global governance frameworks must evolve to oversee transnational data flows.

The refugee ID dilemma highlights the complexity of building systems for people outside of state protections. Digital tools can restore dignity—but only if built with accountability, consent, and the rights of the vulnerable in mind.

Conclusion: Together, these four case studies present a vivid spectrum of digital identity in action—from nation-led platforms to stateless survival tools. They raise critical questions for the future of digital diplomacy and human rights: Who gets to define identity? What systems ensure its protection? And how can we design inclusion in a digital world that is rapidly redefining the boundaries of citizenship?

DIGITAL DIPLOMACY IN THE AGE OF IDENTITY

As digital ID systems proliferate, their implications spill into foreign policy, migration, trade, cybersecurity, and development assistance. Diplomats must now consider:

- How to harmonize digital identity systems across borders while respecting sovereignty.
- How to ensure that international aid programs don't create second-class citizenship through technological gatekeeping.

- How to embed human rights into the architecture of identity technologies.

Diplomatic missions increasingly require secure ID verification for consular services. Trade deals increasingly involve digital trust frameworks. Cyber alliances are shaped not just by threats—but by shared values around data governance.

In this sense, **digital identity has become both a diplomatic asset and a strategic imperative.**

CONCLUSION: TO BE SEEN IS TO BELONG

In an increasingly digital world, recognition is no longer just about being named—it's about being **coded, verified**, and **connected**. Digital identity systems promise inclusion, but they also carry the risk of exclusion for those not seen by the algorithm or not trusted by the state.

The challenge for diplomacy is not merely to regulate technology, but to defend dignity. To ensure that identity systems empower rather than control. That every person—stateless or citizen, migrant or minister—has the right to be recognized not only in a database, but as a full participant in humanity's digital future.

Because in the end, digital identity is not about systems. It's about people. And people deserve to be more than entries in a ledger. They deserve to belong.

Chapter 15: The Impact of Technology on Human Rights Advocacy

"In the digital age, the defense of human dignity depends as much on connectivity as on courage."

INTRODUCTION: A NEW DIGITAL FRONTLINE FOR HUMAN RIGHTS

Technology has fundamentally altered the landscape of human rights advocacy. In a world where violations can be captured, shared, and condemned in real time, the digital age has become both a battlefield and a beacon for justice. Across continents, digital platforms have enabled marginalized voices to rise, authoritarian abuses to be exposed, and global coalitions to form with unprecedented speed.

Yet, this same digital ecosystem also enables:
- State surveillance of activists
- Algorithmic discrimination

- Digital censorship and erasure
- Weaponized disinformation

The digital age offers more than new tools—it offers a new terrain, where data replaces **testimony, algorithms influence access, and visibility becomes protection or vulnerability.**

TECHNOLOGY AS A TOOL OF LIBERATION AND OPPRESSION

Empowering Human Rights Advocacy:
- **Real-Time Witnessing:** Smartphones turn ordinary citizens into documentarians, capturing police abuse, war crimes, or forced evictions as they happen.
- **Global Mobilization:** Hashtag activism (e.g., #BlackLivesMatter, #MeToo, #JusticeForMahsa) connects local struggles to global audiences, spurring cross-border pressure.
- **Data for Justice:** Satellite imagery, forensic analysis, and blockchain-based verification are used to track deforestation, expose illegal mining, and preserve chain of custody for war crimes.

- **Digital Organizing:** Messaging apps and encrypted networks allow movements to self-organize without traditional infrastructure.
- **Counter-Narratives:** Marginalized communities reclaim their stories through podcasts, web platforms, and online documentaries.

Repressive and Dangerous Uses of Technology:

- **Authoritarian Tech-Stack:** Governments deploy surveillance drones, facial recognition, social scoring systems, and spyware to track and crush dissent.
- **Censorship Infrastructure:** Internet shutdowns, deep packet inspection, and geo-blocking are used to silence protests and erase dissent from digital memory.
- **Psychological Warfare:** Troll farms, harassment bots, and targeted disinformation campaigns aim to discredit, intimidate, and traumatize activists.
- **AI-Driven Discrimination:** Predictive policing and algorithmic profiling replicate and intensify existing racial, gender, and class-based inequalities.
- **Digital Colonialism:** Corporate platforms extract data from vulnerable populations without consent,

often without offering meaningful access, benefit, or recourse.

STRUCTURAL CHALLENGES IN THE DIGITAL HUMAN RIGHTS ECOSYSTEM

1. Unequal Access and Infrastructure

- Many communities lack stable internet, digital literacy, or access to safe platforms.
- Language exclusion in major platforms marginalizes non-English speakers from discourse and defense.

2. Platform Governance Gaps

- Content moderation policies are inconsistently applied, often lacking transparency or accountability.
- Platform responses to hate speech or threats vary wildly by region, often under-protecting Global South users.

3. Lack of Legal Protections

- Most international human rights frameworks have not yet fully integrated digital rights.
- Few states have laws regulating spyware, AI bias, or the use of biometric data in law enforcement.

4. Weak Diplomatic Leverage

- Existing human rights mechanisms (UNHRC, ICC, regional bodies) struggle to investigate or act on violations rooted in tech ecosystems.

EXPANDED CASE STUDIES IN TECH AND HUMAN RIGHTS

Case Study 1: The Rohingya Genocide and the Role of Satellite Evidence

In the dim early hours of dawn, the silence of Rakhine State in western Myanmar was pierced by the sound of gunfire, the crackle of flames, and the anguished cries of families being driven from their homes. Between 2016 and 2018, more than 700,000 Rohingya Muslims fled Myanmar to escape what the United Nations would later describe as "a textbook example of ethnic cleansing." But while the government sealed off access to journalists and human rights observers, it could not hide from the digital eye in the sky.

With no reporters allowed in, no mobile networks in place, and a military determined to erase both people and their history, documentation of these crimes fell to an unexpected witness: satellite imagery. Organizations like Amnesty International, Human Rights Watch, and Planet Labs launched an urgent effort to capture daily snapshots of Rohingya villages from space. Their satellites did not blink. They captured the before-and-after images of entire communities, showing village layouts one day and charred nothingness the next. In some frames, fresh military installations rose directly on top of what had once been Rohingya homes—an unmistakable sign of erasure and replacement.

But it wasn't just overhead surveillance that told the story. On the ground, survivors used Facebook, WhatsApp, and mobile cameras to share horrifying testimonies and footage of mass graves, bodies in rivers, and families torn apart. Many posted videos with geotags, which investigators later cross-checked with satellite photos. Though fuzzy and pixelated, these images became powerful tools of verification. They exposed the chronology of violence in a way that eyewitnesses alone could not.

This digital record would become a foundation for legal action. In 2019, the Gambia brought a case before the International Court of Justice (ICJ), accusing Myanmar of genocide. The evidence included satellite images, time-coded videos, and expert forensic reports derived from open-source intelligence (OSINT). The visuals were so compelling that judges ordered Myanmar to implement emergency measures to protect remaining Rohingya populations.

But while the technology provided clarity, the path to justice remained clouded. Political will stalled sanctions. Accountability was limited. Aung San Suu Kyi, once a global icon of peace, defended the military in court. As of today, no top generals have been tried in international criminal courts.

Key Takeaways:

- **OSINT Breakthrough:** This case marked one of the first times satellite imagery and digital content served as core evidence in a genocide case.
- **Democratized Evidence Collection:** Survivors were able to document atrocities without needing traditional media access.
- **Limits of Technology:** While data exposed truth, it could not compel action on its own—highlighting the limits of digital documentation without political enforcement mechanisms.

Case Study 2: #EndSARS and Decentralized Digital Activism in Nigeria

In October 2020, the streets of Lagos, Abuja, and cities across Nigeria throbbed with chants of "End SARS!"—a cry that began online and became a national awakening. The target was SARS—the Special Anti-Robbery Squad—an elite police unit long accused of brutality, extortion, and murder. For decades, Nigerians had feared encounters with SARS officers, many of whom operated with impunity. But it wasn't until smartphones and social media became common tools in the hands of everyday citizens that a tipping point arrived.

A video surfaced on Twitter, allegedly showing SARS officers dragging two young men out of a hotel and shooting one in cold blood. It spread like wildfire, fueling rage and heartbreak. Within hours, hashtags like #EndSARS, #StopKillingUs, and #ReformPoliceNG trended globally. Young Nigerians—many in their twenties—mobilized both online and on the ground. Protest locations were announced on Instagram stories. Legal aid and medical volunteers were coordinated via Google Docs. Donations flooded in from the diaspora—channeled through Bitcoin to evade government freezes on local bank accounts.

This was activism without a central leader, and that was its strength. Decentralization made the movement harder to suppress. However, the Nigerian government soon responded with force. On October 20, soldiers opened fire on unarmed protesters at the Lekki Toll Gate in Lagos. The livestreamed horror—blood on the pavement, screams amid gunshots, people singing the national anthem while dying—seared itself into the global conscience. The government denied the massacre, but evidence uploaded in real time made it impossible to erase.

Despite the mass mobilization, SARS was only formally "disbanded" on paper. No top commanders were prosecuted. In the months that followed, protesters were harassed, travel bans were issued, and bank accounts frozen. Still, a shift had occurred. Millions of Nigerians—young, tech-savvy, and fearless—had found their political voice.

Key Takeaways:

- **Digital Fluency as Power:** Activists turned social media, livestreams, and decentralized coordination tools into weapons of civic resistance.
- **Financial Innovation:** Bitcoin enabled the movement to bypass financial repression, offering a case study in protest finance.
- **Fragile Victories:** While the movement created global awareness and forced police reform rhetoric, structural change remains elusive.

Case Study 3: The Pegasus Spyware Scandal and Abuse of State Surveillance

It began quietly—just a trickle of reports that smartphones were acting strangely, draining battery life faster than usual or turning on microphones without user prompts. But in 2021, an international consortium of journalists shattered the silence with a bombshell: the Pegasus Project. Their investigation revealed that governments across the globe—both authoritarian and democratic—were using a military-grade spyware tool called Pegasus, developed by Israel's NSO Group, to spy on journalists, activists, lawyers, and even heads of state.

Pegasus didn't require a click. It could infiltrate phones silently, accessing everything from photos and text messages to encrypted apps like Signal and WhatsApp. Once inside, it could turn the device into a live surveillance tool—tracking location, recording conversations, and filming through the camera. The targets weren't just criminals or terrorists. They included Mexican journalists investigating cartel ties to politicians, Indian activists supporting minority rights, and French President Emmanuel Macron.

Most haunting of all were the targets close to Jamal Khashoggi—the Washington Post columnist who was brutally murdered in the Saudi consulate in Istanbul. Forensic analysts found Pegasus on the phones of his fiancée and close associates, suggesting they were under surveillance before and after his killing.

The diplomatic fallout was swift. The United States blacklisted NSO Group. France opened a criminal investigation. The United Nations called for a global moratorium on spyware technology. But legal accountability proved elusive. NSO claimed it only sold the software to vetted governments for "counterterrorism." Yet the lines between safety and suppression had been irreparably blurred.

Pegasus exposed a terrifying truth: digital authoritarianism is now for sale. There are no borders in this war—only vulnerabilities. And while private companies profit, civil society bears the cost.

Key Takeaways:

- **Weaponized Surveillance:** Pegasus demonstrated how smartphones could become tools of repression without a trace.
- **Lack of Oversight:** There is no international treaty governing spyware trade—allowing abuse to flourish in legal grey zones.
- **Urgent Call for Regulation:** The case highlighted the need for global digital rights frameworks and transparent governance of surveillance technologies.

RECOMMENDATIONS FOR A HUMAN RIGHTS-BASED DIGITAL ORDER

For States and International Institutions:

1. Codify Digital Rights as Human Rights. Update UN frameworks (ICCPR, UDHR) to include data protection, algorithmic fairness, and freedom from digital surveillance.

2. Establish International Oversight for Surveillance Tech. Create export controls, ethics review boards, and licensing systems for spyware and biometric tools.

3. Protect Digital Civic Spaces. Support internet freedom infrastructure (VPNs, circumvention tools) and protect whistleblowers and online journalists.

4. Reform Multilateral Human Rights Mechanisms. Add dedicated digital rights rapporteurs and enforcement mandates in human rights councils.

For Civil Society and Activists:

1. Invest in Digital Security Training. Build communities of practice around encryption, anti-surveillance, and data hygiene.

2. Develop Decentralized Archival Systems. Use blockchain and distributed storage to preserve documentation in the face of takedowns.

3. Build Cross-Movement Digital Solidarity. Share tools, playbooks, and infrastructure across global movements (e.g., Hong Kong, Sudan, Colombia).

For Tech Companies and Developers:

1. Adopt Human Rights Impact Assessments (HRIAs). Evaluate all products for potential misuse in surveillance or repression.

2. Ensure Algorithmic Transparency. Open-source content moderation models and enable third-party audits.

3. Prioritize Local Language Support. Protect marginalized users by hiring culturally fluent content reviewers and expanding non-Western languages.

CONCLUSION:
JUSTICE IN THE DIGITAL AGE

We are living in an era where truth can be livestreamed—or scrubbed. Where resistance is organized in DMs—but tracked by malware. Where visibility is both protection and threat.

The question is not whether technology will shape human rights—but **who gets to shape the technology.**

We must fight for a future where:
- Platforms protect the vulnerable.
- Tools serve justice, not surveillance.
- Dissent is encrypted—not erased.

The digital age demands a new kind of advocacy—**hybrid, ethical, networked, and relentless.**

It is time to reclaim technology as a force not for domination, but for dignity.

>Not for control, but for connection.

>Not for silence, but for **solidarity.**

Chapter 16: Digital Ethics and Responsibilities

TECHNOLOGY CHANGES THE TOOLS. ETHICS DEFINES THE BOUNDARIES

In the digital age, we are no longer asking whether we can do something—we are now constantly challenged to ask whether we should.

The integration of digital technologies into diplomacy has unlocked vast possibilities: real-time crisis response, direct citizen engagement, cross-border coalition-building, and transparent storytelling. Yet alongside these advances come difficult ethical questions—about privacy, power, manipulation, equity, and truth.

When information moves faster than accountability, and when algorithms influence perception as much as facts do, the ethical dimension of digital diplomacy becomes not just important, but foundational.

This chapter explores the key ethical considerations facing diplomats, policymakers, technologists, and civil society actors operating in the digital sphere. It argues that ethical decision-making is no longer a supplement to strategic planning—it is its core safeguard.

THE ETHICAL IMPERATIVE IN DIGITAL DIPLOMACY

Digital diplomacy sits at the intersection of power, technology, and trust. Every action—whether sending a tweet, launching a platform, or releasing data—carries ethical weight.

Key ethical dilemmas include:

- **Truth vs. Influence:** Can strategic messaging ever justify the omission of facts?
- **Access vs. Exploitation:** Are we building inclusive tools, or extracting attention from vulnerable populations?
- **Speed vs. Deliberation:** In the name of urgency, are we bypassing necessary ethical reflection?
- **Visibility vs. Consent:** When spotlighting suffering, do we empower or further exploit?

Practicing digital diplomacy ethically requires more than compliance. It demands **moral clarity, humility, and foresight**.

CORE PRINCIPLES OF DIGITAL ETHICS IN DIPLOMACY

1. Transparency. Stakeholders should understand how and why digital decisions are made. Algorithms, data use, and moderation must be auditable and explainable.

2. Accountability. Diplomatic actors must take responsibility for the digital tools, messages, and platforms they deploy—even when outsourced.

3. Informed Consent. Whether gathering data or amplifying voices, ensure subjects know how their information or stories will be used.

4. Equity. Digital policies and platforms must not reinforce bias, exclusion, or cultural erasure. Representation and access matter deeply.

5. Non-Manipulation. Diplomatic messaging must avoid psychological coercion, emotional engineering, or algorithmic manipulation—especially of vulnerable groups.

EXPANDED CASE STUDIES: ETHICS IN ACTION AND OMISSION

Case Study 1: Cambridge Analytica When Data Became a Weapon

In the spring of 2018, a quiet unease rippled through the global political and tech spheres. A whistleblower named Christopher Wylie, with pink hair and a calm demeanor, stood before television cameras and testified to a truth many had feared but few had fully grasped: our digital selves had been weaponized.

Cambridge Analytica, a UK-based political consulting firm, had harvested the personal data of over 80 million Facebook users—without their knowledge or consent. Through a seemingly innocent personality quiz app, the firm not only collected data from participants but accessed the friend networks of each user, amassing vast datasets that could model behavioral patterns, emotional triggers, and political leanings.

This wasn't just about targeting voters by region or age. It was about psychological microtargeting—sending individuals messages that tapped into their specific fears, grievances, or aspirations. If a user was deemed neurotic and anxious, they might receive fear-based immigration ads. If they were conservative but disengaged, they'd be pushed conspiracy-laden stories meant to drive them to the polls.

What made this different from traditional campaigning was its invisibility. Most people had no idea they were being profiled, manipulated, or nudged. Political discourse didn't unfold in town halls or on televised debates—but in private, personalized echo chambers, crafted by algorithms and hidden from public scrutiny.

Facebook's delayed and tepid response sparked global outrage. The platform had known about the misuse for years but did little to notify users or prevent further data exploitation. The result was a global crisis of trust—not only in social media, but in democracy itself.

Diplomatic and Ethical Consequences:
- **Democratic trust was shaken**, especially in the U.S. and UK, where the 2016 elections and Brexit referendum were both tainted by revelations of psychological manipulation.
- **Global regulators—particularly in the EU—launched investigations**, prompting the passage

and enforcement of data protection laws like GDPR.

- **The debate over "digital sovereignty" gained traction**, as countries began demanding stricter controls over how foreign tech firms handled their citizens' data.

Key Lessons:

- **Consent is non-negotiable** in democratic societies—data extracted without knowledge is a violation of dignity and autonomy.

- **Ethics must precede innovation**—technological capability should never outpace moral responsibility.

- **Digital diplomacy must advocate for rights-first governance**—or risk being co-opted by surveillance capitalism.

Case Study 2: The Ethiopian Civil War Moderation and the Price of Invisibility

In the hills and valleys of Tigray, where ancient churches cling to cliffs and generations have lived in relative seclusion, a modern war erupted in 2020. But while bullets flew on the ground, a different kind of violence spread online—one built on hashtags, doctored videos, and posts laced with hate.

As the civil war between Ethiopian federal forces and the Tigray People's Liberation Front (TPLF) intensified, social media became a battlefield of its own. Platforms like Facebook and TikTok filled with incendiary content—calls to ethnic violence, fabricated footage, and slurs aimed at demonizing entire communities. These posts were not fringe—they were viral, algorithmically boosted by systems designed to prioritize engagement over accuracy.

The problem was not just content—but language. Amharic and Tigrinya, two of Ethiopia's most widely spoken languages, were virtually invisible to Facebook's moderation systems. Automated filters were built to detect hate in English and a handful of global tongues—not in the languages of Ethiopia. Civil society organizations had warned Meta (Facebook's parent company) about this risk long before the conflict spiraled—but no meaningful steps were taken.

As a result, hate speech spread unchecked. Ethnic tensions deepened. Violence offline mirrored the violence online. People were killed not just with guns—but with rumors.

The United Nations and international watchdogs condemned the platforms. But for many Ethiopians, especially those in the diaspora, it felt like a painful reminder: their lives, languages, and safety mattered less to the global tech ecosystem.

Diplomatic and Structural Implications:

- **Tech firms were accused of "digital colonialism"**—deploying products without adequate support or accountability in the Global South.

- **States began calling for localized moderation systems and stronger accountability** frameworks, particularly in conflict-prone regions.

- **The case exposed the deep biases within AI**, which prioritizes high-resource languages while leaving billions unprotected.

Key Lessons:

- **Language equity is an ethical imperative.** AI must serve all humanity—not just the digitally privileged.

- **Proactive engagement with local civil society is essential** for ethical platform governance.

- **Tech diplomacy must advocate for universal digital rights**, especially in fragile states where misinformation can be deadly.

Case Study 3: Ukraine Ethics of War Storytelling in the Digital Age

When Russian tanks crossed into Ukraine in February 2022, the world watched in real time. But unlike wars of the past, this one unfolded not only on battlefields—but on smartphones, Instagram feeds, and TikTok videos. Ukrainian civilians became frontline correspondents. Drone footage of destroyed villages, tearful goodbyes at train stations, and live broadcasts from bomb shelters flooded the internet.

The digital visibility of this war was staggering. It created an unprecedented window into the human cost of conflict—one that stirred solidarity across continents. Donations poured in. Protests erupted. Governments pledged support. But along with this awareness came difficult ethical questions.

Some videos captured the last moments of a person's life. Others showed grieving children or mutilated bodies. Shared often without consent, they raised the uncomfortable question: where is the line between truth-telling and exploitation? In one now-iconic clip, a grandmother stood defiantly before a Russian soldier, handing him sunflower seeds "so flowers would grow where he died." It became a symbol of resistance. But other images—more raw, more painful—risked reducing trauma to clickbait.

Ukrainian officials used social media not just to inform, but to shape global narratives. President Zelenskyy, in particular, became a master of digital wartime diplomacy—addressing parliaments and people alike through carefully crafted video appeals. His messages humanized the Ukrainian struggle—but always walked a tightrope between emotional persuasion and ethical restraint.

Meanwhile, some foreign media outlets reposted civilian footage without attribution or consent. Well-meaning viewers, in sharing graphic content, sometimes retraumatized survivors or exposed locations to Russian targeting.

Diplomatic and Ethical Balancing Act:

- **Storytelling became both a weapon and a shield**, allowing Ukraine to win hearts and influence policy—but requiring careful curation to avoid turning victims into spectacles.

- **International media were forced to reconsider their ethics**—especially around consent, trauma, and re-use of user-generated war content.

- **Diplomacy had to evolve to become trauma-informed**, emotionally intelligent, and deeply respectful of those it aimed to defend.

Key Lessons:

- **Visibility is powerful—but not inherently ethical.** The right to be seen must be balanced with the right to grieve in dignity.

- **Consent must guide content sharing**, especially in wartime.

- **Diplomats and communicators must develop digital empathy**, combining strategic messaging with respect for human pain.

NAVIGATING THE GREY AREAS: KEY ETHICAL TENSIONS

Deepfakes for Good? Can synthetic media ever be used ethically in advocacy or diplomacy if labeled transparently?

Countering Disinformation with Psychological Nudges: Should diplomats use behavioral science to influence opinions, or does that verge on manipulation?

Content Removal vs. Censorship: When does fighting hate speech become silencing dissent?

These aren't problems to be solved once. They are **tensions to be managed continually**—with evolving norms, contextual judgment, and inclusive input.

RECOMMENDATIONS FOR ETHICAL DIGITAL PRACTICE IN DIPLOMACY

1. Create Ethics-by-Design Protocols
Design digital strategies with built-in ethical reviews, similar to legal or security assessments.

2. Establish Ethics Advisory Boards
Include philosophers, technologists, trauma experts, and civil society in ongoing ethical governance.

3. Mandate Informed Consent in Storytelling
When highlighting victims or vulnerable communities, seek informed consent and clarify use.

4. Build Global Digital Rights Frameworks
Support international treaties on digital privacy, surveillance limits, AI transparency, and online harm prevention.

5. Train for Ethical Foresight
Diplomatic training should include scenario planning that addresses future ethical dilemmas, not just today's known risks.

CONCLUSION: ETHICS AS THE GUARDRAIL OF POWER

In the past, ethics in diplomacy may have seemed like a moral ideal—important, but secondary to realpolitik. In the digital era, ethics is no longer optional. It is the **guardrail that keeps innovation from becoming exploitation.**

Every click, every codebase, every campaign carries with it the power to uplift or to harm. Practicing diplomacy in a digital world means making constant choices—about whose truth is heard, whose lives are seen, and whose dignity is preserved.

Digital diplomacy will define the next generation of global relationships. Its ethical foundation will define whether those relationships are built on trust—or broken by indifference.

The tools are powerful.

The stakes are global.

The compass must be **moral**.

Chapter 17: The Battle for Truth

POWER IN THE DITIAL AGE IS NOT ABOUT REACH – IT IS ABOUT RESTRAINT

Digital diplomacy has unlocked powerful new tools for statecraft—real-time communication, broad public engagement, data-driven insight, and global narrative shaping. But with each new capability comes a deeper question: Just because we can, should we?

As diplomats engage more visibly and more often in digital spaces, the ethical boundaries of their work are being tested. Platforms that amplify voices can also spread harm. Data that offers insights can also invade privacy. Algorithms designed to connect can just as easily mislead.

In this chapter, we examine the ethical dilemmas, governance gaps, and accountability challenges in the emerging field of digital diplomacy—and explore how states, institutions, and individuals can lead with principle, not just power.

WHY DIGITAL DIPLOMACY REQUIRES A NEW ETHICAL FRAMEWORK

Diplomatic behavior has long been shaped by formal codes: the Vienna Convention, bilateral treaties, norms of decorum and discretion. But digital diplomacy moves faster than legislation, and often beyond jurisdiction. This means:

- There are **no universal rules** on how diplomats should use social media.
- There are **few agreed norms** on state behavior in digital engagement.
- The **consequences of digital missteps**—misinformation, reputational damage, international incidents—can be immediate and irreversible.

As digital tools become extensions of diplomatic power, they must also become subjects of **ethical scrutiny and global governance.**

FIVE ETHICAL CHALLENGES FACING DIGITAL DIPLOMACY

1. Disinformation and Digital Weaponization

Should diplomats engage in covert influence campaigns? What are the ethical limits of countering propaganda?

While diplomacy traditionally seeks influence, digital disinformation—whether state-sponsored or outsourced—undermines public trust. When states adopt manipulative tactics, even for strategic gain, they erode the very legitimacy of diplomacy.

2. Privacy vs. Intelligence

Digital diplomacy often relies on open-source intelligence, sentiment analysis, and real-time data collection. But at what point does public listening become surveillance? And how do we protect the privacy of citizens—especially in crisis zones or authoritarian regimes?

Ethical digital diplomacy must find a line between understanding public mood and violating civil liberties.

3. Representation and Digital Inclusion

Whose voice is heard in digital diplomacy? Urban elites with fast internet? English-speaking publics? When foreign policy is shaped by online engagement, diplomats risk missing vast communities who are offline, censored, or digitally excluded.

A just digital diplomacy model must proactively seek out underrepresented perspectives, not just the loudest hashtags.

4. Platform Accountability

Diplomats often depend on commercial platforms—owned by private tech giants—to reach global audiences. But what happens when those platforms:

- Amplify harmful content?
- Are banned in certain regions?
- Allow state-backed actors to harass or distort?

Do diplomats have a duty to push for platform reform—or to find alternatives? The **intersection of public diplomacy and corporate governance** is increasingly a space of friction and opportunity.

5. Institutional Transparency

When a digital misstep occurs—whether a misleading post or a failed online campaign—who is held accountable? Many ministries lack clear internal protocols for digital ethics. And in some cases, digital diplomacy is outsourced entirely to consultants, PR firms, or automated systems.

Institutions must build internal **ethical governance frameworks** that align with foreign policy goals, human rights principles, and national values.

CASE STUDY 1: THE ETHICS OF MEMES IN WARTIME – UKRAINE AND THE WEAPONIZATION OF HUMOR

When Russian troops invaded Ukraine in February 2022, many expected tanks, bombs, and headlines. Few anticipated that memes—yes, internet memes—would become part of the defense strategy.

Ukraine's digital resistance was as multifaceted as its military one. On TikTok, young Ukrainians posted satirical videos of farmers towing away abandoned Russian tanks. On Twitter, images of badly drawn maps mocked Russian propaganda. Even President

Zelenskyy participated in this digital theater, releasing selfie-style updates from Kyiv that often carried dry wit and defiant humor. One image of a Ukrainian soldier flipping the middle finger at a Russian drone circulated globally with the simple caption: "Welcome to Ukraine."

Memes, in this context, were not just jokes. They were weapons—tools to sustain morale, win hearts abroad, and humanize a brutal conflict in ways that resonated across generational and cultural lines. Humor gave people relief amid despair. For international observers, especially younger audiences often detached from traditional news media, these viral posts made the war real, urgent, and emotionally relatable.

But the very nature of humor is double-edged. Critics quickly emerged, arguing that jokes about Russian casualties or mocking the incompetence of invading forces risked trivializing real suffering. Could a meme about a burned tank inadvertently mock the young conscript who died inside it? Did satire blur the line between righteous resistance and gleeful dehumanization?

Some analysts warned that wartime humor—especially in meme form—can foster echo chambers, polarize global debates, and desensitize people to the horror of violence. In a conflict where truth is already under assault, memes, though powerful, could easily become tools of manipulation rather than illumination.

The Ukrainian case revealed that digital diplomacy is no longer limited to communiqués and carefully worded statements. It lives in GIFs, hashtags, and viral humor. But with that power comes ethical responsibility.

Key Reflections:

- **Morale vs. Manipulation:** Humor can sustain spirits and global attention, but may also obscure pain or reinforce stereotypes.

- **Consent and Consequence:** Unlike formal communications, memes often use images and themes without permission or full context.

- **Emotion Must Serve Purpose:** Emotional engagement is critical—but in digital diplomacy, it must be guided by empathy, not sensationalism.

Takeaway

Digital diplomacy walks a fine line between engagement and exploitation. In times of war, humor can unify and uplift—but also desensitize and divide. The true test lies in using emotion not to inflame, but to connect—with dignity and humanity.

CASE STUDY 2: STATE-CONTROLLED NARRATIVES – THE SAUDI ARABIA EXAMPLE

Saudi Arabia has long invested in shaping its international image. Through glittering economic summits, strategic partnerships, and influencers flown in to post selfies in Riyadh, the kingdom has made digital diplomacy a cornerstone of its global branding. But beneath the surface, a darker digital undercurrent has raised serious ethical alarms.

In 2018, the brutal murder of Washington Post journalist Jamal Khashoggi inside the Saudi consulate in Istanbul shocked the world. The murder wasn't just physical—it was digital. In the months and years leading up to his death, Khashoggi had been the target of relentless harassment by Saudi-linked troll armies. Bot networks flooded his posts with threats, spread disinformation about him, and tracked his online activity.

Investigations revealed the Saudi state had orchestrated massive disinformation campaigns, targeting dissidents not just at home, but across the diaspora—from Canada to Turkey to the United States. These weren't isolated incidents. Reports uncovered coordinated online attacks on women's rights activists, critics of the monarchy, and even foreign journalists covering human rights in the region.

At the heart of this strategy was a twisted redefinition of digital diplomacy. The same tools that nations use to build mutual understanding—social media, influencers, real-time engagement—were being turned inward to intimidate, discredit, and silence.

Saudi officials justified their actions as defending the national image against "foreign slander." But the line between reputation management and suppression had long been crossed. When digital campaigns are used to dehumanize critics, stifle dissent, and manipulate global opinion, the cost is not just reputational—it is ethical and diplomatic.

Key Reflections:

- **Narrative Defense vs. Narrative Control:** While states have a right to present their perspective, digital harassment violates international norms.

- **Global Spillover:** Cyber intimidation is borderless. Digital repression of dissidents abroad represents a new form of transnational authoritarianism.

- **Digital Ethics Are Not Optional:** States that engage in digital diplomacy must be held to the same human rights standards as they are offline.

Takeaway

Digital diplomacy is not a shield for authoritarianism. When states weaponize digital tools to silence criticism and intimidate opposition, they erode the very legitimacy they seek to project. In the digital age, soft power must also be ethical power.

CASE STUDY 3: THE EUROPEAN UNION'S "CODE OF PRACTICE ON DISINFORMATION"

While many countries grapple with digital chaos reactively, the European Union has taken a more proactive path—one grounded in transparency, accountability, and values-driven governance.

The EU's "Code of Practice on Disinformation" is a landmark effort to regulate how tech platforms handle misleading or manipulative content. Voluntarily signed by companies like Meta, Google, and TikTok, the Code commits them to a set of ethical behaviors: disclosing the sources of political ads, cracking down on fake accounts and bot networks, promoting reliable journalism, and collaborating with fact-checkers.

More than just a policy document, the Code is a diplomatic framework. It allows states, companies, and civil society to work together to confront the complex challenge of disinformation—without undermining free speech. It also empowers EU diplomats to speak with moral authority on global digital issues, positioning the Union as a leader in ethical tech governance.

This model matters. In an era where disinformation is often deployed as a geopolitical tool—undermining elections, sowing chaos, and deepening division—the EU's approach provides a blueprint for integrity. It shows that digital diplomacy is not merely about countering narratives; it's about building the systems and partnerships that uphold truth itself.

Key Reflections:

- **Voluntary, but Impactful:** Though not legally binding, the Code has spurred real changes in platform policy and cross-border collaboration.

- **Trust-Building Through Transparency:** The EU's framework fosters public confidence by making platforms more accountable.

- **A Values-Driven Approach:** By centering rights, inclusion, and transparency, the EU asserts a diplomatic identity rooted in democratic ethics.

Takeaway

Ethical digital governance is diplomacy in action. The EU's Code of Practice demonstrates that integrity can be institutionalized—not just through regulation, but through shared commitment. In the information age, trust is the strongest currency.

BUILDING A GLOBAL GOVERNANCE ARCHITECTURE

To ensure digital diplomacy advances the global good, the international community must pursue coordinated ethical frameworks. This includes:

- **Norm-building at the UN**, especially through forums like the OEWG and UNGGE on responsible state behavior in cyberspace.

- **Creating a digital diplomacy charter**, akin to the Vienna Convention, that outlines professional conduct in digital spaces.

- **Institutional ethics boards** within ministries to evaluate digital campaigns and manage risk.

- **Transparency requirements** for digital diplomatic interventions, such as watermarking official posts and reporting algorithmic targeting.

Ethics must become an operating system—not just a feature—of digital diplomacy.

CONCLUSION: THE RESPONSIBILITY TO LEAD WITH INTEGRITY

In the digital world, diplomacy is not only about managing relationships—it's about shaping norms. Every post, campaign, and online interaction reflects not just a message, but a method. And in an age of deepfakes, disinformation, and digital division, the greatest power a diplomat holds may be the **power to act ethically when no one is watching—yet everyone is affected**.

Governance begins with example. Ethical leadership in digital diplomacy is about asking the hard questions:

- Will this post build trust or fuel division?
- Does this message empower people or manipulate them?
- Am I using this tool, or is it shaping me?

In the absence of rules, values must guide. And in the presence of complexity, character must lead.

As we look to the future, the next chapter considers where all of these elements converge: in the **creation of new alliances and digital coalitions** that will define not only how we connect—but who gets to shape the world that's coming.

Part IV: Digital Power and Sovereignty

Chapter 18: Blockchain and Decentralized Governance

"In a world where trust in institutions is eroding, blockchain offers not just a new technology – but a new way to imagine power, participation, and transparency."

INTRODUCTION: TRUST, BROKEN AND REBUILT IN CODE

Across the globe, institutions face a growing crisis of legitimacy. Governments struggle to maintain the confidence of their citizens. Financial systems teeter under the weight of inequality. Elections are questioned, contracts are broken, and centralized authorities are often seen not as guardians—but as gatekeepers of privilege.

Into this void stepped an unexpected answer: not a new ideology, but a new architecture. A structure built not of laws or speeches, but of math and consensus. **Blockchain technology**—a decentralized ledger that verifies truth without requiring trust in a central authority—has inspired thinkers, technologists, and diplomats to ask: what if governance itself could be reimagined?

Though first made famous by cryptocurrencies like Bitcoin, blockchain's potential reaches far beyond finance. It offers a new model of participation, where **transparency is built-in, corruption becomes traceable, and communities can make decisions without intermediaries.**

This chapter explores how blockchain is not just disrupting markets—it's challenging centuries-old concepts of power and sovereignty, while offering bold new tools for diplomacy, development, and democracy.

FROM CURRENCY TO CONSENSUS: WHAT BLOCKCHAIN REALLY OFFERS

At its core, blockchain is a way to record information across a distributed network of participants, ensuring that every action—whether a payment, a vote, or a contractual agreement—is visible, verifiable, and immutable.

But its deeper promise lies in what it removes: **the need for trust in a central authority.** In a blockchain system, trust is shifted from institutions to protocols. This raises radical possibilities:

- Citizens could vote without fear of manipulation.
- Aid could be distributed transparently, with no middlemen.
- Diplomatic agreements could self-execute based on verified actions.

Yet for all its promise, blockchain also brings complexity. The technology is often misunderstood. Its ecosystem is rife with hype, scams, and regulatory uncertainty. And while decentralization is its ethos, implementation still requires real-world governance—and human responsibility.

So how does this tension play out in practice?

CASE STUDY 1: ESTONIA—WHERE BLOCKCHAIN MEETS NATIONAL GOVERNANCE

In the forests and cobbled streets of Tallinn, an extraordinary digital experiment has quietly reshaped what it means to govern in the 21st century. Estonia, a nation of just 1.3 million people, has become the world's most digitally advanced society—not through splashy headlines or grand speeches, but through deliberate, radical rethinking of the state itself.

When Estonia regained independence from the Soviet Union in 1991, it inherited no legacy bureaucracy, no entrenched systems—just a clean slate and the urgent need to rebuild. Unlike many post-Soviet states that rushed to reconstruct 20th-century governments, Estonia asked a different question: what if the government could be reimagined as a secure, digital platform?

That vision materialized over three decades. Today, Estonian citizens can file taxes in under five minutes, check medical records online, vote from anywhere in the world, register companies in minutes, and digitally sign any official document. Paper is practically obsolete.

At the core of this system is a blockchain-inspired technology called KSI (Keyless Signature Infrastructure). Developed by the Estonian company Guardtime, KSI doesn't store data but secures it. Every digital action in Estonia's government—from changes to land records to court filings—is cryptographically signed, timestamped, and logged. Any tampering would be immediately visible. This is not just cybersecurity—it's systemic integrity by design.

The stakes became clear in 2007 when Estonia was hit by a wave of Russian cyberattacks, paralyzing banks, media, and government sites. Rather than retreating, Estonia responded with innovation. It became the first country to open a "data embassy" in Luxembourg—a secure offsite storage facility for critical national data, anchoring sovereignty in the cloud.

What makes Estonia's example so powerful is its blend of pragmatism and principle. Blockchain here is not about hype or decentralization for its own sake. It's about trust—codified, embedded, and enforced at scale.

Strategic Insights:
- **Digital society is not just about access—it's about architecture.** Estonia didn't just digitize bureaucracy; it rewired governance itself.

- **Resilience is built on transparency.** Blockchain-like infrastructure has made Estonia one of the most cyber-resilient nations on earth.
- **Trust is a design choice.** The Estonian experience shows that public confidence in government can be engineered through secure, accountable systems.

Takeaway

Estonia's digital governance model offers a powerful example of how blockchain principles—immutability, verifiability, transparency—can rebuild trust in government. It proves that national sovereignty in the digital age isn't about walls—it's about code.

CASE STUDY 2: BLOCKCHAIN FOR HUMANITARIAN AID

In the sweltering heat of Jordan's Azraq refugee camp, where over 30,000 displaced people sought shelter from the war in Syria, an unusual technology quietly changed the way aid was delivered.

Launched in 2017 by the World Food Programme (WFP), the "Building Blocks" initiative replaced paper vouchers and third-party banks with blockchain. Refugees, many of whom had no passports or bank accounts, accessed food by scanning their irises at biometric terminals installed in local supermarkets. Each transaction was recorded on a private blockchain—secure, transparent, and immediate.

There were no middlemen. No bank delays. No risk of misallocated funds. For the first time, families had control over what they received, when, and how. For the WFP, overhead costs fell sharply, fraud was nearly eliminated, and distribution became more efficient.

But beyond logistics, something deeper was happening. For many refugees, this was their first encounter with digital identity. It wasn't tied to a state or a nationality—but to their own biometric data. In a world where over 850 million people lack legal identity, this offered a glimpse of a future where inclusion isn't defined by borders or paperwork.

Yet questions emerged. Who controls this data? Could it be misused? Do the refugees have a say in how the system works? What happens if they leave the camp—or the blockchain fails? And can technology really empower, if it's built and governed by institutions the users can't influence?

Despite the complexities, the Azraq experiment marked a breakthrough. It showed that blockchain can do more than power finance or speculation—it can restore dignity, autonomy, and efficiency where traditional systems have failed.

Ethical and Strategic Reflections:

- **Efficiency without accountability is not enough.** Refugees must be seen not just as beneficiaries, but as stakeholders.

- **Blockchain can democratize aid—but only if human rights are built into its design.**

- **Digital identity is both a lifeline and a vulnerability.** Protecting it must be paramount.

Takeaway

Blockchain in humanitarian aid offers hope—but also a test. It demands that we rethink not only how we deliver help, but how we share power. Technology must not just bypass corruption—it must build equity.

CASE STUDY 3: DECENTRALIZED AUTONOMOUS ORGANIZATIONS AND THE DIPLOMACY OF CODE

In late 2021, the internet erupted with enthusiasm as thousands of strangers came together to buy a rare, original copy of the U.S. Constitution. They weren't part of a political party, a government, or a nonprofit. They were part of ConstitutionDAO—a decentralized autonomous organization formed entirely online.

In less than a week, over $40 million in cryptocurrency had been pooled through smart contracts and shared governance protocols. Members—from artists in Berlin to engineers in Seoul—had equal say. They voted on wallet strategy, legal representation, and auction tactics. Though they ultimately lost the bid, the effort made headlines worldwide.

But the deeper story was not about one document—it was about a new form of organization.

DAOs function like internet-native collectives. They don't have CEOs or headquarters. They operate through code—smart contracts that execute decisions automatically once consensus is reached. This structure promises radical transparency, borderless cooperation, and distributed power.

The potential is vast. Some DAOs have funded climate projects, coordinated disaster relief, launched media platforms, and even explored participation in multilateral governance efforts. They are experimenting with new models of diplomacy—ones not bound by geography, but by shared mission and programmable trust.

Yet, for all their promise, DAOs inhabit a legal and ethical gray zone. Who is accountable if a DAO is hacked? Can it be sued? Can smart contracts truly represent human complexity—compromise, emotion, context?

Some warn that DAOs risk becoming ideological silos or playgrounds for tech elites. Others see them as laboratories for a more participatory future—where diplomacy emerges not just from foreign ministries, but from communities, algorithms, and code.

Governance and Ethical Reflections:

- **DAOs challenge traditional diplomacy**, forcing institutions to grapple with new forms of digital political will.

- **They offer transparency, but also risk volatility.** Their success depends on code—but also on the human communities behind it.

- **Legal recognition is still murky.** Until states recognize and regulate DAOs, their diplomatic influence remains aspirational.

Takeaway

DAOs may not replace governments—but they are redefining who gets to organize, advocate, and act in the global arena. In the diplomacy of the future, power might no longer be held solely by states—but by networks governed by code, transparency, and shared purpose.

THE DIPLOMATIC IMPLICATIONS OF BLOCKCHAIN GOVERNANCE

As blockchain systems mature, they are beginning to intersect with global diplomacy in unexpected ways. Nations now negotiate over cryptocurrency regulations. International institutions explore digital identity frameworks. Activists and whistleblowers use decentralized tools to protect speech and organize movements.

This shift raises urgent diplomatic questions:
- Can treaties be encoded as smart contracts?
- How do we hold DAOs accountable across jurisdictions?
- Can blockchain infrastructure itself become a neutral ground for peacebuilding?

And beyond policy lies a deeper cultural transformation. Diplomats accustomed to formal protocols and hierarchies must now engage with communities that prize **transparency, speed, and decentralization**. This calls for new literacies—not just technical, but ethical, participatory, and narrative.

CONCLUSION: DESIGNING TRUST FOR A FRAGMENTED WORLD

The world is facing a crisis not only of governance, but of trust. Traditional institutions—governments, banks, media—are viewed with skepticism. In this context, blockchain does not promise perfection. It promises a

new possibility: systems where power is distributed, records are transparent, and decisions are made in the open.

But technology is not destiny. A blockchain can empower a community—or entrench inequality. It can protect rights—or encode injustice. What matters is **who designs the system, who governs it, and who is left out.**

The diplomacy of the future must therefore do more than adopt new tools. It must embrace new values: openness, decentralization, and accountability—not as technical features, but **as foundations for a more equitable world.**

Blockchain won't fix our politics. But it might help us imagine better ones. And in an age of uncertainty, imagination may be the most radical diplomacy of all.

Chapter 19: The Role of Digital Infrastructure in Development and Sovereignty

"Before a nation can innovate, it must connect. Before it can compete, it must compute."

INTRODUCTION: INFRASTRUCTURE IS IDENTITY

In the 20th century, the mark of a modern nation was its roads, railways, and ports. These physical arteries carried people, ideas, and goods—and with them, the promise of prosperity and sovereignty.

In the 21st century, a different kind of infrastructure is defining power. It is made of fiber-optic cables, data centers, 5G networks, cloud architecture, satellite constellations, and national digital platforms. This is digital infrastructure—the invisible scaffolding that now holds up everything from healthcare systems to diplomatic negotiations, from public services to military defense.

Without it, nations risk falling into **technological dependency**—relying on foreign platforms for essential functions, exposing themselves to surveillance, cyberattacks, and economic exclusion.

But with it—when built thoughtfully and inclusively—digital infrastructure can become the foundation of a **new kind of sovereignty**, one rooted not just in borders or GDP, but in **resilience, access, and global agency**.

This chapter explores how countries are using digital infrastructure not just to modernize, but to reclaim their future.

CASE STUDY 1: RWANDA—BUILDING A DIGITAL BACKBONE IN A POST-CONFLICT NATION

In 1994, Rwanda stood on the edge of collapse. A brutal genocide had claimed the lives of over 800,000 people in just 100 days. Infrastructure was destroyed, institutions decimated, and social trust shattered.

Rebuilding the country was not just about roads or laws—it was about healing a nation from the ground up.

Against this backdrop of trauma, Rwanda embarked on one of the most remarkable development journeys of the 21st century. Under the steady leadership of President Paul Kagame, the country adopted a radical vision: rather than replicate outdated bureaucratic models, Rwanda would build a nation grounded in digital inclusion, innovation, and transparency.

The result was Vision 2020, followed by Vision 2050—comprehensive national blueprints that placed digital infrastructure at the center of Rwanda's development. This wasn't just modernization for its own sake. It was about reimagining what a post-conflict state could be.

One of the first steps was laying a 4,500-kilometer national fiber-optic backbone, reaching even the country's most remote villages. This physical network was followed by digital access: the launch of Irembo, a government e-service portal offering over 100 services—like registering for marriage, applying for passports, or paying traffic fines—without bribes, queues, or red tape.

Rwanda didn't stop at government services. Through its Smart Kigali initiative, the capital city became a testbed for urban connectivity—offering free Wi-Fi on buses, in public parks, and in schools. The nation partnered with Zipline, a drone delivery company, to transport blood to rural clinics. It collaborated with Carnegie Mellon University to establish a graduate program in information and communications technology (ICT), nurturing local talent for the digital age.

More than mere efficiency, these systems were a form of dignity infrastructure. For Rwandans, accessing healthcare without a bribe, enrolling in university from a rural village, or receiving blood by drone wasn't just impressive—it was transformational. It signaled a new relationship between citizen and state: one based on service, not suspicion.

Today, Rwanda positions itself as a pan-African digital hub. It hosts international tech conferences, invests in startups, and promotes regional digital cooperation. The scars of genocide are not erased, but Rwanda's digital future has become a beacon of possibility.

Strategic Insights:

- **Digital tools can rebuild civic trust.** Transparency in service delivery fosters inclusion and dignity.

- **Connectivity must reach beyond capitals.** Rwanda's rural fiber strategy shows the value of national cohesion.

- **Technology is a peacebuilding tool.** When used ethically, digital infrastructure can unite, rather than divide.

Takeaway

Rwanda's digital transformation is not a tech success story—it's a human one. In a nation once torn apart, fiber-optic cables have become threads of unity, linking past trauma to future hope.

CASE STUDY 2: UNITED ARAB EMIRATES FROM OIL TO OPTIC FIBER

Fifty years ago, the United Arab Emirates was a sparse desert federation defined by pearl diving, caravans, and coastal trade. Today, it's home to AI-powered police stations, blockchain-powered government services, and autonomous vehicle testbeds. How did this transformation happen so quickly—and so deliberately?

The answer lies in the UAE's understanding that infrastructure is power—not just in the traditional sense, but in its digital form. With vast oil reserves and global trade revenue, the Emirates had the wealth to sit back. Instead, it chose to reinvest that capital into its own digital reinvention.

This journey took form through visionary policy. The Smart Dubai initiative, launched in 2013, promised to make Dubai the "happiest city on Earth" through seamless digital governance. And it wasn't just a slogan. In practice, it meant:

- **Blockchain integration across all government functions**, making transactions secure and transparent;
- The rollout of **UAE Pass**, a single national digital identity allowing citizens and residents to access hundreds of services with one login;
- The establishment of the **Mohammed bin Zayed University of Artificial Intelligence**, the world's first graduate-level AI-focused institution;

- Nationwide investment in **IoT, 5G, cloud computing, and autonomous mobility infrastructure**.

But perhaps most striking is how the UAE has turned digital infrastructure into a tool of diplomacy. Hosting regional internet exchange points, satellite gateways, and major hyperscale data centers, the Emirates has positioned itself as the **digital crossroads of the Middle East.** By controlling key routes in the global data economy, the UAE is no longer just a consumer of connectivity—it's a sovereign architect of it.

This evolution is about more than modernization. It's about soft power, branding, and regional leadership. In a world where **data is the new oil**, the Emirates has ensured that its pipelines run not only beneath the ground—but through fiber, clouds, and code.

Strategic Insights:
- **Infrastructure is strategy.** Digital platforms are the new ports and airports of global influence.
- **Data sovereignty is the new geopolitics.** Hosting infrastructure brings power—and responsibility.

- **Smart governance builds legitimacy.** Citizens experience the state not through slogans, but through services.

Takeaway

The UAE's leap from hydrocarbons to hyperscalers illustrates how digital infrastructure can become a new form of statecraft—reshaping global influence through bandwidth, trust, and design.

CASE STUDY 3: KENYA—SILICON SAVANNAH AND THE PROMISE OF KONZA

In Nairobi's bustling streets, where informal markets coexist with tech accelerators and matatus (public minivans) now accept mobile payments, a quiet digital revolution has been unfolding. Kenya, long known for its vibrant entrepreneurial culture, has emerged as a continental tech leader—earning it the nickname "Silicon Savannah."

This rise began with **M-Pesa**, the mobile money platform launched in 2007 that leapfrogged traditional banking and revolutionized how Kenyans pay, save, and send money. But the momentum didn't stop there.

In 2009, Kenya unveiled a bold vision: the creation of **Konza Technopolis**, a master-planned city 60 kilometers south of Nairobi designed to be Africa's Silicon Valley. Part of **Kenya Vision 2030**, the project includes:

- A technology-focused university and research center;
- Tier III data centers for secure digital operations;
- Business parks to host both local startups and multinational tech firms;
- Green smart-city infrastructure, designed around sustainability and efficiency.

Though Konza has progressed more slowly than hoped, its symbolism matters. It is **a monument to ambition**—a statement that Kenya sees its future in innovation, not extraction.

Meanwhile, the rest of the country has not stood still. The government expanded broadband access, launched digital ID programs, and supported dozens of startup incubators and tech hubs, especially in Nairobi. Companies from Google to Microsoft now run major programs in Kenya, drawn by its talent, connectivity, and policy openness.

Still, challenges remain. Debates around **data privacy, platform accountability,** and **rural digital inclusion** are growing. Civil society groups demand that the digital boom must also serve the underserved—those without smartphones, connectivity, or digital literacy.

Strategic Insights:

- **Leapfrogging works—if paired with inclusion.** Skipping legacy systems is an opportunity to build better—but not for the few.
- **Cities reflect identity.** Konza is as much a message as a municipality.
- **Ecosystems must be nourished.** Infrastructure must be matched with education, ethics, and entrepreneurship.

Takeaway

Kenya's digital journey is a negotiation between possibility and equity. It proves that infrastructure alone isn't destiny—but when anchored in vision and participation, it becomes a pathway to a more connected and just future.

THE STRATEGIC STAKES OF INFRASTRUCTURE

Why does infrastructure matter so much?

Because without it, even the most elegant policies and ideas collapse. Digital identity systems fail without secure databases. E-governance platforms crash without reliable bandwidth. Diplomacy flounders if servers are vulnerable to foreign interference.

Infrastructure determines:
- Who gets online—and who stays invisible
- Whose systems are trusted globally—and whose are bypassed
- Whether a country is a digital colony—or a digital leader

But infrastructure is never neutral. It reflects political will, values, and trade-offs. Building a national cloud is not just a technical project—it's a statement about sovereignty, independence, and resilience.

And when infrastructure is outsourced—when countries rely entirely on foreign vendors, platforms, or undersea cables—they may lose more than data. They may lose the **right to govern their own digital destiny.**

CONCLUSION:
BUILDING THE INVISIBLE NATION

The future will be shaped not only by who has the best ideas, but by who builds the best systems to support them. Digital infrastructure is the foundation of diplomacy, democracy, and development in the 21st century. But it must be **built with people in mind**, not just profit margins or geopolitical games.

For some nations, the challenge is catching up. For others, it's leading wisely. For all, the task is urgent: to design infrastructure that reflects **human dignity, sovereign control, and inclusive growth.**

Because in this new era, a strong nation is not only one that controls its land—but one that owns its cloud, secures its code, and uplifts its citizens through a digital infrastructure worthy of their dreams.

Chapter 20: Algorithms at the Table

NEW NETWORKS, NEW NORMS, NEW RULES OF ENGAGEMENT

The old world of multilateral diplomacy was defined by marble halls and multigenerational treaties. It moved slowly but deliberately. Consensus-building was an art, and the weight of history shaped every vote.

But in today's digital age, that world is being rewritten in real time. Now, alliances form not only around territory, defense, or ideology—but around data, platforms, standards, and code. The players have changed. The rules are still being written. And the pace is breakneck.

This chapter explores how digital diplomacy is reshaping multilateralism—through innovative partnerships, emerging coalitions, and new global governance architectures that reflect the needs, tensions, and hopes of a digitally interdependent world.

WHY MULTILATERALISM MUST EVOLVE

Traditional multilateral institutions—like the UN, WTO, and World Bank—were designed for a different era. While they remain central to global cooperation, they often struggle to respond to digital realities:

- Jurisdiction is blurry.
- Tech platforms operate transnationally but aren't states.
- Norms around cybersecurity, data privacy, and AI are fragmented and contested.

What's needed now is a new layer of multilateralism—more agile, inclusive, and responsive to the challenges of our digital future.

THREE FORCES DRIVING DIGITAL MULTILATERALISM

1. Global Interdependence on Infrastructure

Every nation now relies on shared undersea cables, cloud platforms, satellite networks, and semiconductor supply chains. A disruption in one country can affect connectivity across an entire region.

This demands shared stewardship—and collective crisis readiness—for global digital infrastructure.

2. Cross-Border Data Governance

Data does not respect borders, yet countries vary widely in their rules. The tension between data localization, privacy, surveillance, and free flow of information is a geopolitical flashpoint—and a ripe area for multilateral alignment.

3. The AI Race and Tech Power Imbalance

As a handful of countries and corporations develop frontier AI technologies, others are left behind or left vulnerable. Without inclusive governance, AI may amplify inequality and disrupt global stability.

A multilateral response is not just ideal—it is essential.

Case Study 1: The Global Partnership on AI (GPAI)

Diplomacy in the Age of Algorithms

In the early months of 2020, as the world wrestled with a global pandemic and deepening digital divides, a coalition of governments quietly launched an initiative with outsized ambition: to shape the future of artificial intelligence (AI) not just through innovation—but through inclusion, rights, and responsibility.

This was the birth of the **Global Partnership on AI (GPAI)**—a multilateral forum bringing together countries like France, Canada, India, Japan, and members of the European Union. What set GPAI apart from other international tech bodies wasn't just its mandate, but its design: it was **flexible, expert-driven, and value-based.**

Rather than being a treaty organization or regulatory authority, GPAI was conceived as a **policy platform—where governments, academia, civil society, and industry experts** work side by side. Together, they form working groups that focus on pressing themes like AI governance, data justice, innovation ecosystems, and responsible AI in the pandemic response.

At the heart of GPAI is a bold idea: **that technology must serve people—not the other way around.** This means anchoring AI development in **human rights, democratic values**, and **international cooperation**. It also means rejecting a purely economic or militarized lens on AI—a stance that distinguishes GPAI from the power-politics that often dominates global tech diplomacy.

GPAI also puts strong emphasis on **capacity-building for low and middle-income countries**. It recognizes that global inclusion in the AI age requires not just principles, but resources—funding, training, and infrastructure to ensure that the benefits of AI do not remain concentrated in a few hands.

As of today, GPAI remains a work in progress. It has no binding enforcement powers, and its reach is not yet global. But it reflects a new kind of diplomacy—**consensus-based, multi-stakeholder, and ethically grounded.**

Key Highlights:

- GPAI reflects a shift from geopolitics to **principle-based alliances**, showing how international cooperation can center values rather than borders.

- By including **non-Western countries like India**, GPAI broadens the frame of global AI governance beyond traditional blocs.

- Its emphasis on **human rights, transparency, and development equity** marks a clear stand against the weaponization or unchecked commercialization of AI.

Takeaway

The Global Partnership on AI is not just an initiative—it's a blueprint for how we govern emerging technologies with integrity. It suggests that the future of diplomacy may be shaped less by embassies and more by **collaborative, cross-sector working groups** dedicated to shared human ideals in a rapidly changing world.

CASE STUDY 2: The Digital Cooperation Organization (DCO)

Rebalancing the Digital World

In 2020, in the heart of Riyadh, a different kind of multilateral alliance quietly took shape. Unlike most digital governance forums, which are dominated by Western democracies and headquartered in the Global North, this organization emerged from the Global South—with a mission to empower it.

The **Digital Cooperation Organization (DCO)** brings together countries such as **Saudi Arabia, Bahrain, Nigeria, Pakistan, and Rwanda**, united not by geography or ideology, but by a shared goal: to accelerate digital development in emerging economies and ensure their voices are heard in shaping the global tech agenda.

What makes the DCO unique is its structure and speed. It's lightweight, member-led, and action-focused—designed not for grand declarations, but for practical cooperation. Its key priorities include:

- **Digital entrepreneurship:** supporting startups, SMEs, and women-led digital businesses through mentorship, funding, and policy advocacy.

- **Skills development:** providing digital literacy and advanced tech training across youth populations.

- **Infrastructure and access:** investing in internet connectivity, cloud services, and inclusive digital ID systems.

- **Regulatory harmonization:** helping member countries align their digital laws to facilitate cross-border innovation and trade.

The DCO's emergence also reflects a deeper **rebalancing of global tech governance**. As the West grapples with internal regulatory battles and geopolitical rivalries, the DCO offers an alternative voice—one rooted in local realities, demographic opportunity, and economic urgency.

It doesn't aim to replace institutions like the OECD or the G7, but rather to complement them—filling a gap where many lower-income nations feel excluded or underrepresented. And in doing so, it opens up space for a **more pluralistic and culturally attuned digital order.**

Key Highlights:

- The DCO **elevates digital voices from the Global South**, offering them agency in rulemaking, standard-setting, and tech policy dialogue.

- Its structure is **flexible and adaptive**, well-suited for fast-moving digital economies and diverse legal contexts.

- By focusing on **development-led cooperation**, the DCO provides a new model **for digital solidarity** beyond traditional aid paradigms.

Takeaway

The Digital Cooperation Organization shows that digital diplomacy is no longer the sole domain of the Global North. In an increasingly multipolar world, new alliances are forming around shared needs, not historical alliances. The DCO is a quiet but powerful reminder that the future of digital governance must be built with—not just for—**emerging economies**.

CASE STUDY 3: The EU's Digital Decade and Transatlantic Tech Alliance

Values at the Core

In Brussels, the future of technology is not just about innovation—it's about governance. The European Union has positioned itself as a global norm-setter in digital affairs, grounded in the belief that rights, rules, and responsibilities must travel as fast as data itself.

Over the past decade, the EU has passed landmark legislation that now serves as global benchmarks. Chief among them:

- **The General Data Protection Regulation (GDPR)** — a global standard for data privacy and user consent.

- **The Digital Services Act (DSA)** — which imposes accountability on platforms for illegal content and misinformation.

- **The Digital Markets Act (DMA)** — designed to prevent monopolistic behavior among "gatekeeper" tech companies.

Together, these form the heart of the EU's **"Digital Decade"**—a vision for a secure, open, and human-centered digital future.

But the EU's digital diplomacy doesn't stop at its borders. Through the **EU–U.S. Trade and Technology Council (TTC)**, Europe has formed a critical transatlantic alliance with the United States—one that aims to coordinate on:

- Semiconductors and supply chain resilience
- Platform regulation and disinformation
- AI governance and global tech standards
- Countering digital authoritarianism, particularly from China and Russia

The TTC represents a delicate but vital attempt to bridge regulatory differences across the Atlantic. While the U.S. tends toward innovation-first and self-regulation, the EU favors rights-first, institutional frameworks. Finding common ground is both a challenge and a necessity.

Together, this alliance offers a **liberal democratic counterweight** to authoritarian digital models. It is not about imposing uniformity, but about **defining a shared digital ethos** in an era of geopolitical uncertainty.

Key Highlights:

- The EU's approach centers on **rule-of-law, transparency, and user protection**, offering a viable path for democratic digital governance.

- Through TTC, the EU and U.S. are **aligning strategic priorities** to ensure that the digital world reflects open values, not coercive controls.

- This cooperation helps defend against **techno-authoritarianism** by embedding ethical norms into cross-border digital infrastructure.

Takeaway

The EU's digital strategy is diplomacy by regulation. It shows that you don't need to build the most powerful platforms to shape their behavior. Through law, partnership, and principled engagement, the EU and its transatlantic allies are helping to write the rules for a digital world that respects both innovation and integrity.

EMERGING MODELS OF DIGITAL MULTILATERALISM

1. "Mini-Laterals" and Digital Compacts

Small coalitions of like-minded countries (e.g., Quad Tech Working Group, AUKUS Cyber) form around specific issues: quantum computing, cyber defense, semiconductor supply chains.

2. Multistakeholder Governance

Tech platforms, civil society, and NGOs are increasingly included in governance forums—from Internet Governance Forum (IGF) to AI summits. This reflects a **post-Westphalian reality**: power is dispersed, and diplomacy must be shared.

3. Sovereign Tech Alliances

Countries are pooling resources to create regional digital solutions: African data centers, Latin American e-government platforms, Gulf fintech ecosystems. This is **decolonizing digital dependence** and fostering **regional digital resilience**.

RISKS AND CHALLENGES AHEAD

Despite this progress, digital multilateralism faces significant headwinds:

- **Geopolitical fragmentation** between China, the U.S., and the EU on standards.

- **Lack of enforcement mechanisms** for cyber norms and AI ethics.

- **Digital exclusion**, where least-developed countries lack voice or access.

Without inclusive governance, digital diplomacy could reproduce the same inequalities and asymmetries it seeks to solve.

CONCLUSION: FROM COMPETITION TO COOPERATION

Digital alliances are the new frontiers of multilateralism. But they are more than strategy—they are statements of intent. They show what kind of digital world we want to build. One rooted in openness, fairness, sustainability, and peace—or one fragmented by fear, control, and isolation.

The future of diplomacy will be defined by coalitions—not just of states, but of values. The road ahead is not without conflict, but the foundation is being laid. What we decide now—on data, on AI, on platforms, on infrastructure—will echo for generations.

The choice is ours: fragment or federate, compete or cooperate, react or lead.

Digital diplomacy, if done right, doesn't just adapt to the future. It helps create it.

Chapter 21: Global Cooperation in the Digital Age

"In a digitally connected world, global challenges require diplomacy to solve problems and maintain trust in algorithms, platforms, and infrastructure"

A NEW GLOBAL ECOSYSTEM NEEDS NEW DIPLOMACY

As blanketed by cables, servers, satellites, and platforms, the world today operates in a digital ecosystem that knows no sovereignty. A rogue disinformation campaign in one country can influence elections half a world away. A data breach by a global tech company can expose users in countless nations. A virus can cascade across networks before it becomes visible as a public health risk.

Traditional diplomacy—embassies, summits, sealed agreements—is no longer sufficient for such rapid, cross-border complexity. The global digital order demands a different mode of cooperation: one rooted not in trust between states alone, but in collective governance over systems, standards, and shared norms.

This chapter explores how states, intergovernmental bodies, and civil society organizations are shaping such frameworks, often imperfectly, but always urgently—because **digital sovereignty is not a solo endeavor: it is a global pact.**

CASE STUDY 1: THE UN GLOBAL DIGITAL COMPACT—TOWARD SHARED DIGITAL VALUES

In an era when algorithms increasingly shape our lives, and data flows transcend borders more easily than people, the global digital landscape has outpaced our systems of governance. By 2021, it was clear to many within the United Nations that the internet—and the digital tools built upon it—could no longer be governed in fragmented national silos. Something more unified was needed. Something more principled.

The result was the beginning of consultations for the **Global Digital Compact (GDC)**—a proposed international agreement aimed at defining shared principles for the digital age. It was envisioned not as a binding treaty, but as a moral framework to guide digital development around the world. Access, human rights, inclusion, trust, and sustainability were placed at the center of the draft dialogue.

The GDC aims to do what few other agreements have attempted: bring together high-tech powers and low-income countries, private platforms and public agencies, civil society and cybersecurity experts. From Silicon Valley to Sub-Saharan Africa, it seeks to build consensus around what the digital future should look like—and who it should serve.

Among the core principles under discussion:

- **Universal access to affordable, meaningful connectivity**
- **Protection of digital human rights**, including privacy, freedom of expression, and data dignity
- **Promotion of digital inclusion**, particularly for women, marginalized communities, and the Global South

- **Safeguards around AI and emerging technologies**, ensuring they remain human-centered
- **Cybersecurity cooperation**, to prevent the digital realm from becoming a domain of conflict

Of course, such ambition comes with friction. Critics argue the Compact is slow-moving, toothless without enforcement, and potentially undermined by geopolitical competition. Powerful states and platforms may pay lip service while continuing practices that contradict the spirit of the Compact.

But the process itself matters. By creating **space for dialogue across divides**, the UN is planting the seeds for a new kind of diplomacy—one that is participatory, values-based, and anticipatory. This isn't just about catching up with tech; it's about shaping it before it shapes us beyond recognition.

Key Highlights:

- The GDC represents an **ethical foundation** for digital governance, not just technical coordination.
- It redefines global cooperation in the digital sphere as **multilateral and inclusive**, not transactional.

- Its greatest strength is **soft power through consensus**, not enforcement.

Takeaway

The Global Digital Compact may not be a finished product, but it is a sign of a turning point: the digital world must be governed not only by innovation—but by intention. And intention, in the form of shared values, is where future legitimacy will live.

CASE STUDY 2: G7 & G20 DIGITAL INITIATIVES

Aligning Economies Through Shared Norms

For decades, multilateral diplomacy revolved around treaties, tariffs, and summit communiqués. But in the last decade, a new dimension has emerged: the **diplomacy of digital standards**. Here, the G7 and G20 have become key arenas where geopolitics meets gigabytes.

The **G7**, comprising the world's most advanced industrial economies, has increasingly prioritized digital cooperation. Since 2017, its **Digital and Tech Ministers' Meetings** have addressed urgent cross-border challenges—from data localization and cybersecurity to the governance of AI and the regulation of tech giants.

Recent joint communiqués have highlighted commitments to:

- **Coordinated cybersecurity frameworks** across member nations;
- **Global tax policy for digital services**, ensuring fair contribution from tech companies;
- **Combatting online misinformation** and holding platforms accountable for content moderation;
- **Establishing ethical AI guidelines** centered on transparency, accountability, and fairness.

Meanwhile, the **G20**—with a broader membership including emerging powers like India, Brazil, and Indonesia—has complemented these efforts. Its Osaka Track, launched in 2019 under Japan's presidency, sought to promote the **free flow of data with trust**, enabling cross-border e-commerce and cloud infrastructure development.

What distinguishes these efforts is their pragmatism. Rather than pursuing sweeping legal instruments, the

G7 and G20 aim for policy convergence—where shared principles create **alignment without coercion.**

This reflects a subtle, but profound shift in diplomacy: from **hard law to soft coordination**, from domination to mutual stewardship of digital ecosystems. And it signals recognition that in the 21st century, **digital rules shape economic futures as much as trade ones once did.**

Key Highlights:

- G7 and G20 forums offer a platform for **tech diplomacy among equals**, balancing innovation with governance.
- Their initiatives promote **regulatory harmony**, reducing friction for cross-border data and services.
- They reflect **emerging trust architecture**, where cooperation replaces competition in digital rulemaking.

Takeaway

The G7 and G20 digital tracks show that today's most powerful diplomatic tool is not the pen—but the protocol. As these forums coordinate norms in real time, they ensure that economies don't just grow together—but govern together, too.

CASE STUDY 3: ITU AND REGIONAL STANDARDS—CONNECTIVITY AS COMMON GROUND

While the United Nations crafts broad ethical frameworks and the G20 aligns policies among leading economies, another layer of digital diplomacy plays out behind the scenes—among engineers, regulators, and standard-setters. At the center of this world stands the **International Telecommunication Union (ITU)**.

As the UN's oldest specialized agency, the ITU has long played a quiet yet crucial role in ensuring that the **networks connecting us are compatible, secure, and fair**. From frequency allocation for satellites to setting standards for 5G and (soon) 6G technologies, the ITU lays the groundwork for **technical interoperability across borders**.

But in recent years, its role has expanded—from technical facilitator to **diplomatic convener**. The ITU now brings together governments, telecom companies, and tech firms to discuss digital sovereignty, data governance, and the future of global internet architecture.

Recent initiatives include:

- **Global principles for internet governance**, balancing openness with security;
- **Cybersecurity frameworks** shared across continents, particularly in vulnerable states;
- **Satellite and spectrum coordination**, reducing friction in space-based connectivity;
- **Support to regional alliances**—like ASEAN, the African Union, and Mercosur—in aligning digital policies with global norms.

This is diplomacy not through declarations, but through **standardization**. When countries agree on spectrum use or security protocols, they build more than networks—they build trust.

And for many developing nations, **ITU-aligned regional frameworks are lifelines**. They allow smaller economies to plug into a broader digital economy, access funding and expertise, and protect their digital sovereignty without building every system from scratch.

Key Highlights:

- ITU standardization enables **global interoperability**, preventing digital fragmentation.
- Its role as a **technical bridge-builder** makes it essential to inclusive digital development.
- Regional adoption of ITU frameworks helps harmonize governance across borders and industries.

Takeaway

In the digital era, infrastructure is diplomacy—and standards are its language. The ITU proves that behind every network connection lies not just code, but consensus. And in that quiet consensus, global cooperation finds one of its most durable forms.

THE ROLE OF CIVIL SOCIETY AND TECH COMMUNITIES

Importantly, multilateral digital diplomacy is not just between governments. **Tech communities, human rights advocates, standard bodies, and diaspora groups** increasingly participate in policy design.

Organizations like the **World Wide Web Consortium (W3C)** create open standards. NGOs like **AccessNow** and **Civic Tech** advocate for digital rights at UN forums. Global youth movements demand that AI governance honor generational equity.

Digital cooperation platforms like **DiploFoundation** and the **Digital Future Society** act as conveners—bridging official processes and grassroots insight, humanizing policy-making by insisting on narratives, not just protocols.

CHALLENGES: FROM FRAGMENTATION TO FRAGMENTED TRUST

Despite progress, digital multilateralism faces deep obstacles:

- **Trust deficits:** Some states perceive digital rules as veiled tools of influence and refuse to adopt Western-led frameworks.
- **Fragmentation risks:** A fragmented digital order—with competing standards and compliance regimes—becomes a network of "splinternets", threatening interoperability and global commerce.
- **Enforcement gaps:** Without strong mechanisms, agreements can become symbolic—trust gestures rather than actionable treaties.

- **Digital inequality:** Rich countries often hold more policy influence, while Global South states may lack the technical capacity to participate meaningfully.

These challenges highlight that technology diplomacy is not a solved problem—it is an evolving experiment requiring empathy, equity, and imaginative governance.

TOWARD A HUMANIZED, INCLUSIVE DIGITAL DIPLOMACY

Despite setbacks, there are models of hopeful practice:

- **Inclusive negotiation spaces** ensure Global South tech ministries and civil society representatives have voice.
- **Capacity-building partnerships**, where tech engineers from developing nations collaborate on global governance issues.
- **Digital service reciprocity**, such as open-source public platforms shared across countries.
- **Ethical standard labs**, where diverse communities co-design AI principles or content moderation policy.

In this vision, digital diplomacy becomes not just a policy exercise, but a shared act of constructing global solidarity.

CONCLUSION: DIPLOMACY FOR A NETWORKED FUTURE

As the world leans deeper into digital life, multilateral cooperation is no longer optional. If history taught us that peace requires treaties, we must now recognize that **trust requires code**, that sovereignty requires standards, and that human rights require platform governance.

The multilateral diplomacy of the digital era is about more than managing crises—it is about engineering trust, **building shared infrastructure**, and ensuring that the benefits of connectedness are accessible to all, not just a few.

This form of diplomacy will hinge on relationships—not just between states, but between states and platforms; between technologists and civil society; between cities, communities, and institutions. It should be humble, adaptive, and driven by curiosity—not just power.

Because in a world woven together by data, the greatest act of diplomacy may be this: to build systems so just that every person—every human, not just every state—can rely on them without fear.

Chapter 22: The Future of Global Governance in the Digital Age

WHEN RULES LAG BEHIND REALITY, LEADERSHIP MUST LEAP FORWARD

The structures that once held the world together—treaties, international institutions, rules-based order—were designed in an analog age. They were shaped after world wars, not cyberwars. They were built for borders, not bandwidth.

But the world has changed.

Power has shifted.

And digital transformation has pushed global governance to a moment of reckoning.

The future of governance will not be written in ink alone—it will be coded, streamed, encrypted, and hashed. It will be negotiated not just in conference halls, but in cloud servers, virtual summits, and decentralized protocols.

This chapter explores what global governance must become in the digital age: more inclusive, more agile, and deeply human-centered.

THE CRISIS OF TRUST IN TRADITIONAL INSTITUTIONS

Global institutions today are facing a crisis—not just of relevance, but of legitimacy.

The Symptoms

- **Gridlock at the UN Security Council**, paralyzed by geopolitics in the face of humanitarian crises.
- **Outdated trade rules** that fail to regulate e-commerce, digital currencies, or data flows.
- **Weak enforcement mechanisms** on cyber norms, leading to increased attacks with little accountability.

The Deeper Issue

Institutions that were built to stabilize the 20th century are struggling to govern the 21st. They were never designed to manage:

- Global data sovereignty
- Cross-border algorithmic bias
- Platform monopolies
- Artificial intelligence with geopolitical implications

As societies evolve faster than institutions, we must ask: **Who writes the rules for a world where sovereignty, identity, and influence are increasingly digital?**

RETHINKING SOVEREIGNTY IN A DIGITALLY ENTANGLED WORLD

In the digital age, **sovereignty no longer stops at the border.**

A meme launched in Moscow can shape elections in Madrid.

A regulation passed in Brussels can affect code in San Francisco.

A cyberattack in Tehran can shut down a power grid in Toronto.

The digital sphere is not separate from geopolitics—it is now its primary battleground.

This requires a fundamental rethinking of sovereignty, built around:

- **Shared responsibilities** for global infrastructure (like undersea cables, satellites, cloud storage)
- **Joint frameworks** for cross-border data and digital taxation
- **Coordinated cyber defense**, with clear thresholds and norms

GOVERNING EMERGING TECHNOLOGIES—BEFORE THEY GOVERN US

From AI and quantum computing to digital ID and central bank digital currencies (CBDCs), technological innovation is outpacing policy.

The Governance Vacuum

- **AI systems** make decisions that affect millions—but whose ethics are embedded in the code?
- **Digital currencies** challenge traditional banking systems—but where is the consensus on regulation?
- **Biometric data** is collected by both governments and companies—but who owns it?

Global governance in the digital age must move faster, not just reactively. It must be built on principles of:

- **Transparency**
- **Human rights**
- **Cross-sector collaboration**
- **Interoperability across jurisdictions**

TOWARD A NEW ARCHITECTURE OF GLOBAL COOPERATION

To rise to this moment, we need bold institutional innovation.

1. A Digital Bretton Woods Moment

Just as the world once rebuilt the global financial order after WWII, we now need a Digital Bretton Woods— a forum that:

- Sets standards for digital trade, tax, and trust
- Establishes norms for AI safety and ethical use
- Ensures the rights and agency of digitally marginalized nations and populations

2. Inclusive Governance Models

Global governance must no longer be dominated by the same handful of actors. The voices of the Global South, youth, civil society, and indigenous communities must not be "consulted"—they must be co-authors.

The internet is global. Its governance must be, too.

3. Multi-Stakeholder Diplomacy

Diplomacy is no longer only between governments. Tech companies, NGOs, universities, and grassroots movements are now essential to problem-solving.

The future lies in **hybrid institutions**—where public and private actors co-govern with transparency and accountability.

CASE STUDIES IN DIGITAL GOVERNANCE INNOVATION

1. The Paris Call for Trust and Security in Cyberspace. Launched by France in 2018, this initiative brings together governments, civil society, and the private sector to promote responsible behavior in cyberspace.

It's not a treaty. It's a values-based coalition. It includes over 1,000 signatories. And it shows how voluntary, soft law frameworks can move faster than traditional international law.

2. The African Union's Digital Transformation Strategy. Africa is asserting a digital voice on its own terms. The AU's strategy focuses on regional data governance, infrastructure development, and digital inclusion—creating continental sovereignty over its future.

This is not just development. It's diplomacy through empowerment.

3. The UN's Global Digital Compact (in progress). As part of the UN's "Our Common Agenda," the Compact seeks to establish universal principles for digital cooperation—a kind of digital bill of rights for the 21st century.

If successful, it could become the normative anchor for global digital governance.

LEADERSHIP FOR A DIGITALLY GOVERNED FUTURE

To lead in this new era, we don't just need new rules—we need new rulemakers.

Global governance must be guided by moral imagination:

- What does justice look like in a world shaped by algorithms?
- How do we define peace when warfare is hybrid, invisible, and constant?

- What does sovereignty mean when platforms reach further than parliaments?

We need leaders who can hold the weight of tradition in one hand—and the possibilities of innovation in the other.

We need **global stewards**, not just bureaucrats. **Bridge-builders**, not gatekeepers.

CONCLUSION: FROM FRAGILE ORDER TO SHARED DESTINY

The future of global governance in the digital age is still unwritten. But one thing is clear: we cannot copy and paste our way into it. The institutions of tomorrow must reflect **the complexity, connectivity, and collective stakes** of the digital world.

There will be friction. Power will resist reform. And technology will continue to challenge our sense of control.

But there is also hope—because in every corner of the world, leaders, thinkers, and citizens are beginning to ask the right questions. They are stepping forward not with all the answers, but with the **courage to create better ones**.

Let us not simply govern the digital world. Let us govern for it—with wisdom, justice, and humanity at the core.

Because the future will not wait.

And if we rise to meet it, it will not disappoint.

Part V: The Future of Diplomacy

Chapter 23: Education and Training for Digital Diplomacy

PREPARING DIPLOMATS FOR A WORLD THAT UPDATES IN REAL TIME

Diplomacy used to be a craft honed over decades—by observing, by reading between the lines, by mastering the rhythms of negotiation in quiet rooms and ceremonial halls. The tools were consistent. The tempo predictable. A skilled diplomat was one who could deliver a nuanced phrase with perfect timing or draft a communique that could shape regional peace.

But in today's world, that quiet room is gone. The hall has gone hybrid. The communique is now a tweet—subject to virality, misinterpretation, and meme warfare. And the tempo? It's not just fast—it's always on.

In this era of digital transformation, diplomacy can no longer rely solely on tradition. It must be **relearned**,

retooled, and **retaught**. That is why education and training are not simply support functions—they are now the engine of effective diplomacy.

This chapter explores how foreign ministries, universities, international organizations, and technology partners must rethink the training of diplomats—not as a one-time credential, but as a continuous evolution of skill, mindset, and digital fluency.

THE CASE FOR A NEW KIND OF DIPLOMAT

The 21st-century diplomat operates in a world where:

- Crises unfold live on social media.
- Cyber threats challenge national sovereignty.
- Private companies shape international narratives.
- Fintech, AI, and blockchain influence global development.
- Public opinion can shift policy before a memo is written.

In this context, diplomats must be not only linguists and negotiators—but also digital strategists, data interpreters, public communicators, and tech-savvy bridge-builders.

To serve effectively, today's diplomats need interdisciplinary training that spans technology, law, public communication, ethics, international development, and cyber policy.

WHAT DIGITAL DIPLOMATS NEED TO LEARN

1. Digital Literacy and Platform Fluency

Understanding the ecosystem of social media platforms, digital communications tools, and content trends is no longer optional. Diplomats must learn:

- How platforms influence information flows.
- When and how to intervene in digital narratives.
- How to manage disinformation and online reputational risk.
- How to tailor diplomatic messages across media channels.

2. Cybersecurity Awareness and Protocol

While diplomats are not expected to be engineers, they must:

- Know how to safeguard personal and institutional digital security.
- Understand the basics of cyber threat models, attack vectors, and attribution challenges.
- Be able to participate meaningfully in cybersecurity norm-setting discussions.

3. Data Interpretation and Decision-Making

Diplomats now have access to dashboards filled with public sentiment data, real-time analytics, and digital monitoring tools.

Training must include:

- How to interpret sentiment trends in different regions.
- How to make decisions based on incomplete or conflicting data.
- How to combine digital intelligence with cultural context.

4. Technology Policy Fundamentals

From AI ethics to digital identity frameworks, diplomats increasingly participate in global tech governance. They must understand:

- The basics of blockchain, CBDCs, and digital payments.
- International frameworks for data privacy and digital rights.
- The roles of tech companies in public-private diplomatic partnerships.

5. Cross-Cultural Digital Communication

What's persuasive in one online culture can be offensive in another. Diplomats must be trained to:

- Recognize digital cultural norms and sensitivities.
- Manage tone and framing across regions and languages.
- Navigate moments of viral misunderstanding with grace.

HOW TO TRAIN THE DIGITAL DIPLOMAT

1. Diplomatic Academies Must Modernize

Many foreign ministries already operate training academies for incoming diplomats. These institutions must now expand their offerings to include:

- **Simulation labs** for digital crises and cyberattack response.

- **Scenario-based learning** involving social media escalation, digital public diplomacy campaigns, or cyber norm negotiations.

- **Rotations** with tech companies, multilateral digital institutions, or cybersecurity agencies.

2. Partnerships with Universities and Think Tanks

Forward-looking programs are emerging globally, such as Oxford's Digital Diplomacy initiative and DiploFoundation's online programs. Ministries should collaborate with:

- Academic institutions offering interdisciplinary programs in technology and international relations.

- Global think tanks developing policy labs focused on AI governance, cyber norms, and digital trust.

3. In-Service, Ongoing Training

Digital fluency is not static. Platforms, threats, and tools evolve rapidly. Continuous professional development is essential:

- **Refresher courses** on new platforms or security protocols.
- **Workshops** on digital ethics, media literacy, and online engagement.
- **Peer learning groups** to share tactics across missions and regions.

4. Inclusion of Non-Traditional Voices

Effective digital diplomacy requires diverse perspectives. Ministries must create space for:

- **Technologists** who understand digital infrastructure.
- **Younger diplomats** who are digital natives.
- **Civil society and private sector actors** who co-shape the digital space.

CASE STUDY 1: THE UK'S FCDO – INTEGRATING DIGITAL INTO DIPLOMATIC DNA

Training diplomats to think strategically, communicate effectively, and adapt quickly

In the corridors of Whitehall, where foreign policy has traditionally been written in ink and spoken behind closed doors, the UK's Foreign, Commonwealth & Development Office (FCDO) has been quietly rewriting the playbook. Its mission: to embed digital thinking not as an add-on, but as core diplomatic capacity. For Britain, diplomacy in the 21st century isn't just about treaties and telegrams—it's about narratives, networks, and nimbleness.

Background and Motivation

The FCDO realized early that the digital age was not merely transforming communication—it was reshaping geopolitics. From disinformation campaigns to cyberattacks, from viral protests to livestreamed revolutions, the digital arena had become a front line. Recognizing this shift, the FCDO didn't just pivot—it institutionalized change.

Building Skills, Not Just Awareness

In 2015, the UK launched its Diplomatic Academy, designed to prepare a new generation of diplomats for a world where influence is measured in both headlines and hashtags. One of its pillars is the **"Technology and Diplomacy"** module—a comprehensive track where young and mid-career diplomats learn:

- How to engage effectively (and ethically) on social media.

- How to assess cybersecurity threats and advocate for responsible cyber norms.

- How to integrate AI, open-source intelligence, and tech platforms into foreign policy design.

Scenario-based trainings simulate digital crises: misinformation floods during elections, deepfake-driven reputational threats, or real-time online escalations between rival factions. These are no longer theoretical risks—they are the daily realities of diplomacy.

An Open-Source Intelligence Revolution

Another pioneering feature is the FCDO's Open Source Unit, which trains diplomats to harvest actionable insights from publicly available digital data. By analyzing conversations on platforms like Twitter, TikTok, and Telegram, British missions are better equipped to understand shifting sentiments in target regions, detect early signs of unrest, and navigate emerging political narratives.

Real-World Applications

During the COVID-19 pandemic, this approach was stress-tested. As governments scrambled to respond, British embassies used secure cloud platforms to coordinate rapid-response digital campaigns. In regions like Southeast Asia and Sub-Saharan Africa, UK missions produced localized content to combat disinformation, support vaccination efforts, and promote cooperation. All of it was crafted, distributed, and refined using the skills learned at the Academy.

Why This Model Matters

The UK's FCDO shows that digital diplomacy isn't an instinct—it's a skillset. One that can—and must—be taught. Their approach recognizes that diplomats are no longer just envoys; they are strategic communicators, crisis responders, and data-informed analysts.

Most importantly, the UK has demonstrated that **digital literacy must be systemic**—baked into the DNA of institutions, not confined to a few tech-savvy outliers.

CASE STUDY 2: ASEAN'S REGIONAL EFFORTS TO BUILD CYBER AND DIGITAL CAPACITY IN THE GLOBAL SOUTH

In a region as diverse as Southeast Asia—where some countries are global tech hubs and others still face basic connectivity challenges—ASEAN has had to balance ambition with equity. The bloc has chosen a path of solidarity-driven digital diplomacy, where training, collaboration, and trust-building form the backbone of regional progress.

Digital Asymmetries, Shared Goals

From Singapore's hyper-digital infrastructure to the fledgling systems in Laos and Myanmar, ASEAN's member states have dramatically different starting points. Yet all face similar threats: cybercrime, online extremism, digital inequality, and growing dependence on global tech giants.

Recognizing this, ASEAN launched several **flagship capacity-building initiatives:**

- **ASEAN-Singapore Cybersecurity Centre of Excellence (ASCCE):** Based in Singapore, this regional hub offers training on cyber norms, international law, and confidence-building measures. Participants include diplomats, military officials, and civil servants from all 10 member states.

- **ASEAN–Japan Cybersecurity Hub:** Supported by Japan, this initiative delivers technical skills and strategic planning support to emerging ASEAN economies, helping them design and implement national cybersecurity strategies.

- **Digital Diplomacy Labs** with partners like the EU and Australia offer hands-on sessions in digital rights, data sovereignty, cross-border regulation, and inclusive internet governance.

Changing the Face of Cyber Diplomacy
These programs have had measurable impact. For example:

- **Lao and Cambodian diplomats**, once sidelined in regional tech negotiations, are now actively participating in international dialogues such as the UN Open-ended Working Group (OEWG) on cybersecurity.

- ASEAN has adopted **joint statements** in UN cyber forums, allowing the bloc to punch above its weight and influence the global discourse on digital peace and cooperation.

Why This Effort Matters

ASEAN's work illustrates that regional **unity can compensate for national disparities**. It reinforces a central tenet of digital diplomacy: **no country, however small or under-resourced, should be left behind**.

This peer-to-peer, partnership-based model redefines diplomacy from the Global South—not as reactive, but as proactive, collaborative, and increasingly strategic.

CASE STUDY 3: DIPLOFOUNDATION – BUILDING A GLOBAL CADRE OF DIGITAL DIPLOMATS

Training diplomats from every corner of the world – one virtual cohort at a time

For small states and developing nations, access to elite foreign service academies or high-end digital tools is often limited. But since 2002, DiploFoundation has been quietly leveling the playing field—training thousands of diplomats worldwide in the tools, values, and strategies of digital diplomacy.

A Mission Rooted in Access

Founded in Malta and Geneva, DiploFoundation's goal is simple but radical: make diplomacy more inclusive by giving diplomats from all nations, especially underrepresented ones, the training they need to engage in 21st-century global governance.

A Training Model Built for the Digital Age

Courses are conducted online, allowing participation from remote Pacific islands, landlocked African states, and even conflict-affected regions. The curriculum is rigorous and comprehensive:

- Cybersecurity policy and global governance
- AI and ethics in international relations
- Internet governance, misinformation, and cross-border data regulation
- Simulated negotiations on digital trade, misinformation crises, and cyber incidents

Impact Through Inclusion

What makes Diplo unique isn't just its global reach—it's the **tailored approach**. Participants receive context-sensitive education adapted to their political, cultural, and institutional realities.

The results speak for themselves:

- A female diplomat from the Pacific became the first from her nation to **co-author an international policy paper on AI ethics**.
- Several African diplomats trained by Diplo have since gone on to **design national data governance strategies**, with technical support from the World Bank and UNDP.
- Small island nations, often absent from global AI, cybersecurity, and e-commerce talks, are now

active voices in international forums thanks to Diplo alumni.

Why It Matters

DiploFoundation proves that digital diplomacy doesn't require Silicon Valley budgets or Brussels bureaucracies. What it requires is access, empathy, and adaptability. It's a model where **knowledge is a bridge**, and where even the smallest voice can shape international norms.

By equipping diplomats from every continent with practical tools, Diplo shows that the digital future is not just for the powerful—it belongs to **everyone**.

CONCLUSION: LEARNING TO LEAD IN A DIGITAL WORLD

In an age when diplomacy is shaped by tweets, cyber threats, and algorithm-driven narratives, the role of the diplomat is being fundamentally transformed. What was once a career of quiet negotiation behind closed doors is now a vocation lived—often—on an open, fast-moving digital stage. To meet this moment, diplomacy must not only evolve—it must be taught anew.

The case studies in this chapter offer a compelling roadmap. The United Kingdom's Diplomatic Academy demonstrates how institutional modernization can embed digital agility at the core of foreign service. ASEAN's regional efforts show how collaborative, capacity-building training can empower developing states to shape international cyber norms. DiploFoundation proves that diplomacy's digital future can be inclusive, with even the smallest nations able to contribute meaningfully—if given access to the right tools and training.

These examples reinforce a vital truth: **education is no longer a back-office support to diplomacy—it is the front line**. The digital diplomat must be part communicator, part strategist, part technologist—and fully grounded in the ethical and geopolitical complexities of a networked world.

Yet digital diplomacy training is not just about acquiring skills. It's about cultivating **judgment in ambiguity, clarity in crisis, and integrity in influence**. It's about preparing diplomats not only to navigate disruption—but to lead through it.

To do so, we must build systems that:

- Adapt in real time, just as technologies do.

- Include a diversity of voices, from digital natives to underrepresented states.

- Bridge the public and private, combining statecraft with platform expertise.

If we fail to equip diplomats with the tools, insight, and resilience for a digital age, we risk leaving international relations at the mercy of actors who understand technology—but not responsibility. But if we succeed, we empower a generation of global leaders ready to secure peace, promote trust, and solve shared challenges—in a language the world now understands: digital.

As we turn the page, we move from preparation to practice—where stakes are highest, time is scarce, and reputations are forged or lost in moments. In the digital age, crisis diplomacy is where theory meets decisive action.

Chapter 24: Recommendations for Practitioners and Policymakers

BRIDGING THE GAP BETWEEN INNOVATION AND IMPLEMENTATION

Diplomacy is in a remarkable state of transition. We're not just seeing changes in how diplomacy looks; there's a shift in how it functions. Technology is reshaping the landscape of international engagement, paving the way for more transparency, participation, and responsiveness. However, these advancements bring about significant new challenges, including disinformation, cyber threats, digital inequality, and ethical dilemmas.

Today, practitioners and policymakers are tasked with two key challenges: they must adapt traditional diplomatic principles to fit these new realities, while also shaping the frameworks that will define the future of digital diplomacy for generations to come.

In this chapter, we offer practical, experience-based recommendations to guide decision-makers, foreign service professionals, and institutional leaders as they navigate this evolving landscape.

BUILDING DIGITAL FLUENCY WITHIN DIPLOMATIC CORPS

1. Make Digital Literacy Mandatory

Every diplomat, regardless of their experience or location, should have basic training in:

- Social media strategy and risks
- Cybersecurity fundamentals
- Data governance and privacy standards
- The influence of AI and algorithms

2. Create Hybrid Teams

Encourage collaboration between traditional diplomats and technologists, data analysts, behavioral scientists, and digital ethicists. Multidisciplinary diplomacy is resilient diplomacy.

INSTITUTIONALIZE DIGITAL CRISIS RESPONSE

3. Develop Digital Rapid Response Units
Form teams that can monitor, assess, and counter digital threats—like disinformation and viral crises—in real time. Establish clear protocols for escalation and coordinated public messaging.

4. Integrate OSINT (Open Source Intelligence) Tools
Take advantage of publicly available data to proactively track geopolitical tensions, election interference, and online narratives, particularly in fragile regions.

STRENGTHEN CYBERSECURITY AS A DIPLOMATIC PRIORITY

5. Establish Cyber Norms Through Multilateral Channels
Support ongoing discussions at the UN and regional organizations to define:

- Acceptable behaviors in cyberspace
- Critical red lines (e.g., attacks on hospitals or electoral systems)

- Protocols for incident attribution and response

6. Promote Cyber Diplomacy Officers
Appoint specialized envoys to spearhead cyber dialogues, engage with tech platforms, and contribute to global agreements on digital sovereignty and infrastructure protection.

ENGAGE NON-STATE ACTORS STRATEGICALLY AND TRANSPARENTLY

7. Include Civil Society in Norm-Setting
Involve NGOs, researchers, and advocacy groups in international discussions—they provide vital local context, especially on issues like surveillance and privacy.

8. Partner with Tech Companies Proactively
Shift from reactive regulation to collaborative governance. Establish ongoing dialogues with tech platforms, co-create moderation standards during crises, and explore tech diplomacy fellowships.

HUMANIZE DIGITAL POLICY

9. Center Human Rights in Digital Frameworks

Every policy should be assessed from the perspective of:

- Digital inclusion and access
- Freedom of expression
- Protection of marginalized communities
- The psychological impacts of misinformation and online harm

10. Protect Human Agency in AI-Driven Diplomacy

Ensure that humans remain accountable for:

- Automated decision-making in areas like migration and aid distribution
- The ethical use of predictive technologies
- Minimizing bias in diplomatic analytics and language tools

RETHINK COMMUNICATION AND STRATEGIC MESSAGING

11. Shift from Control to Conversation

Today's public diplomacy should prioritize dialogue instead of monologues. Practitioners ought to:

- Engage in real-time discussions
- Acknowledge when there's uncertainty or error
- Empower missions to tailor messaging to local contexts

12. Practice Narrative Resilience

Train diplomats not only to rebut disinformation but also to tell proactive stories that foster trust and preempt manipulation.

ELEVATE ETHICS AND TRUST AS STRATEGIC ASSETS

13. Create Internal Ethics Committees

Diplomatic institutions should have dedicated teams to:

- Review contentious communications
- Guide AI deployment in foreign policy

- Establish ethical guidelines for influence operations

14. Lead by Example in Digital Conduct

Foreign services should model the behaviors they wish to see around the world. Trust is built through consistent and accountable actions.

EXPAND GLOBAL INCLUSION AND CAPACITY BUILDING

15. Support Digital Diplomacy in the Global South

Many regions lack the infrastructure, training, or representation to engage fully in digital diplomacy. Invest in:

- Digital education for young diplomats
- Cyber capacity-building initiatives
- Ensuring equitable representation in global tech governance

16. Create South–South and Triangular Partnerships

Encourage cross-regional learning among emerging digital leaders to share experiences and craft context-aware solutions.

INSTITUTIONALIZE FORESIGHT AND FUTURE-READINESS

17. Establish Digital Diplomacy Foresight Units

These units should monitor emerging technologies—like quantum computing and decentralized governance—and assess their long-term diplomatic implications.

18. Simulate Future Diplomatic Scenarios

Use war-gaming and scenario planning to experiment with future crises and test policy readiness before real consequences arise.

REAFFIRM THE HUMAN CORE OF DIPLOMACY

19. Balance Speed with Deliberation

While digital tools often encourage urgency, diplomacy still requires thoughtful consideration. Allow time for slow thinking in fast-paced environments.

20. Preserve Empathy in Policy

Remember that behind every policy is a person affected. Digital transformation should not come at the expense of the core human values that underpin diplomacy.

CONCLUSION: FROM ADAPTATION TO STEWARDSHIP

The digital age demands more than reaction—it calls for reimagination.

Practitioners must move from adapting to change to shaping it, and policymakers must lead with clarity, courage, and care. The tools may evolve, but the core mission of diplomacy endures: building understanding, preventing conflict, and advancing the collective good.

In a world of shifting algorithms and competing narratives, those who succeed will not simply master the new tools—but will **craft institutions, relationships, and values that outlast them.**

This is not just a technological shift.

It is a **diplomatic renaissance**.

And it begins with intention.

Chapter 25: Reflections on Digital Diplomacy: Embracing Change for a Secure Future

"Digital diplomacy is not just about pixels and protocols – it's about people, purpose, and the preservation of peace"

THE JOURNEY THUS FAR

Over the course of this book, we have traversed a vast landscape of ideas, actors, platforms, and ethical dilemmas that define diplomacy in the digital age. What once was a practice grounded in discreet meetings and sealed communiqués has become a **real-time, transparent, and participatory exercise.**

The digital transformation of diplomacy has reshaped every element of how states and non-state actors engage. Artificial intelligence influences negotiations, social media platforms shape public perception of conflict, and cyber threats can disrupt the foundations of entire societies. Diplomacy has moved from marble corridors to mobile screens—and in doing so, it has become **faster, more exposed, more contested, and more consequential.**

But with this transformation comes a question we must all ask:

Are we shaping the digital world—or is it shaping us?

UNDERSTANDING THE SHIFT: FROM LEGACY TO LIQUID DIPLOMACY

The traditional model of diplomacy was designed for a world of scarcity—of access, voice, and information. It operated within hierarchies and rituals, valuing patience, discretion, and institutional continuity. But the 21st century introduced a different reality:

- **Voice is abundant.**
- **Information is instant.**
- **Narrative is power.**
- **Access is democratized—but uneven.**

We are witnessing a transition from legacy diplomacy to liquid diplomacy—where actors, allegiances, and communication flows are fluid, multidimensional, and increasingly mediated by digital ecosystems.

In this environment, diplomacy must shed outdated assumptions and embrace agility, responsiveness, and co-creation as core competencies. It must become less about formality—and more about function and impact.

A WORLD REDEFINED: DIGITAL REALITIES AND GLOBAL FRAGILITIES

Digital diplomacy does not exist in a vacuum. It is unfolding alongside tectonic global shifts:

- The erosion of trust in institutions and media
- The rise of nationalism and digital authoritarianism
- Climate-induced migration and humanitarian crises
- Growing inequality in digital access and literacy
- An increasingly multipolar world with emerging powers and plural identities

In this context, digital diplomacy cannot merely automate the past. It must **rewrite the rulebook**, while holding fast to enduring values: dignity, peace, dialogue, and mutual respect.

THE FIVE PILLARS OF A SECURE DIGITAL DIPLOMATIC FUTURE

To meet the challenges ahead, digital diplomacy must be built on five foundational pillars:

1. Trust-Centered Engagement

Without trust, digital tools become megaphones for manipulation. Diplomacy must be:

- Transparent in intent
- Consistent in voice
- Respectful of privacy and agency

Trust must also be earned not only from other states—but from global citizens, who increasingly influence diplomatic outcomes through activism, voting, or platform behavior.

2. Ethics-Driven Technology

The tools we build reflect our values. Digital diplomacy must:

- Adopt "ethics by design" principles in algorithm deployment
- Prevent digital colonialism and extractive surveillance
- Ensure human rights frameworks are baked into code

Technology must serve humanity—not the other way around.

3. Resilience Against Digital Threats

Cybersecurity is now national security. Diplomats must:

- Engage in collective cyber defense frameworks
- Establish norms for peaceful conduct in cyberspace
- Create early-warning systems for digital escalation

The digital battlefield is already here—resilience requires foresight, partnerships, and rapid response.

4. Cultural and Emotional Intelligence

Facts alone don't change minds. Diplomacy must:

- Understand digital culture, memes, humor, and narrative codes
- Speak not only to logic, but to identity and belonging
- Promote inclusive storytelling that recognizes marginalized histories and perspectives

This is how diplomacy wins hearts—not just minds.

5. Equity and Inclusion

The digital world mirrors the inequalities of the physical one. We must ensure:

- **Digital literacy and infrastructure access** for all
- Multilingual and culturally adapted content
- Representation of the Global South in AI, content moderation, and cyber norms creation

No digital future can be secure if it is not shared.

EMBRACING COMPLEXITY, LEADING WITH HUMILITY

The future of diplomacy will be shaped by crises we cannot yet foresee—biotechnological risks, AI-driven conflicts, or climate-triggered digital migration.

To succeed in this world, diplomats and decision-makers must:

- **Welcome ambiguity**, rather than fear it
- **Lead with humility**, recognizing that no one institution holds all the answers
- **Build coalitions**, not empires—working across sectors, borders, and identities
- **Invest in learning** as a lifelong diplomatic function

Diplomacy must now think in decades, act in minutes, and feel in human terms.

A PEOPLE'S DIPLOMACY FOR A DIGITAL PLANET

For centuries, diplomacy was the domain of elites. Today, it is increasingly shaped by:

- Online youth movements
- Citizen science and civic tech
- Indigenous data sovereignty initiatives
- Cultural diplomacy driven by creators, not ministries

This is not a threat to diplomacy. It is its **renewal**.

When more people are at the table, diplomacy becomes:

- More **inclusive**
- More **innovative**
- More **accountable**

We must embrace a diplomacy that does not just speak for people—but **with them**.

A FINAL WORD: THE DIPLOMATS WE MUST BECOME

The future diplomat must be a strategist and an ethicist.

A negotiator and a coder.

A storyteller and a systems thinker.

They must see diplomacy not as an act of preservation—but as a **practice of imagination**.

Because digital diplomacy is not just about harnessing new tools.

It is about embracing a new role:

- As guardians of peace in times of digital chaos
- As architects of global trust in the age of division
- As navigators of a future that demands both courage and compassion

The task ahead is vast. The tools are imperfect. The terrain is unstable.

But the mission—**to build a safer, more just, and more human world—remains unchanged**.

And for that, diplomacy is more essential than ever.

Chapter 26: Conclusion

NAVIGATING THE DIGITAL HORIZON – A DIPLOMATIC RECKONING

At the dawn of the 21st century, diplomacy stood at a crossroads. The familiar world of handshakes, treaties, and formal declarations began to blur, as messages traveled at the speed of light, secrets unraveled through leaks, and influence campaigns played out in trending hashtags rather than in marble halls.

Through the chapters of this book, we have witnessed a profound transformation. Not just of tools—but of purpose, posture, and power. The digital revolution has not merely updated diplomacy. It has upended it.

Where once borders defined jurisdiction, now **data flows define influence.** Where embassies once held the secrets of states, now cloud platforms hold the memories of nations. Where diplomacy once relied on protocol, today it requires presence—in the timelines, screens, and algorithms that shape perception and reality.

But amid this change, one truth remains: **diplomacy is still human at its core**. It is the craft of listening, the art of negotiation, and the will to seek peace when conflict is easier. The medium has changed. The mission has not.

WHAT WE'VE LEARNED: THE PILLARS OF DIGITAL DIPLOMACY

From our exploration of cyber diplomacy and artificial intelligence to digital identity and blockchain governance, to inclusion, gender, climate, and crisis communication, several insights emerge as essential:

1. Trust is the New Currency. In an age of misinformation, surveillance, and digital harm, trust—earned through transparency, participation, and ethical governance—has become the lifeblood of diplomacy.

2. Infrastructure is Power. Nations that build and control their digital infrastructure—data centers, platforms, identity systems—define their own sovereignty. Those that do not, risk becoming dependent or vulnerable.

3. Inclusion is Not Optional. Any diplomacy that excludes voices—whether by gender, geography, or bandwidth—undermines its legitimacy. A digital order that doesn't reflect the world's diversity will fail the world's majority.

4. Multilateralism is Imperative. No country can govern the digital realm alone. Cooperation—messy, slow, imperfect—is the only path to collective security and ethical progress.

5. Diplomacy Must Be Ethical, Adaptive, and Imaginative. The speed of innovation will outpace legislation. The only way to lead is to embed values into systems before crises arise—to anticipate harm, and to prioritize dignity from the design phase forward.

LOOKING AHEAD: THE DIPLOMAT OF THE FUTURE

Tomorrow's diplomat will not just speak multiple languages—but understand multiple platforms. They will be as comfortable in policy briefings as in code audits. They will know how to decode algorithms, interpret cultural cues in digital communities, and navigate new domains of sovereignty: the metaverse, cyberspace, and artificial intelligence.

They will need to be part peacebuilder, part ethicist, part technologist. But above all, they will need to be **deeply human**—capable of empathy in an era of automation and guided by purpose in a time of noise.

Because no matter how far technology evolves, it will still be people—navigating power, pain, and possibility—who must lead.

THE FINAL THOUGHT: WIRED, BUT NOT BOUND

We live in an era of wired alliances—networks of connection that stretch beyond treaties and trade routes. These alliances are forged not just between states, but between societies, ideas, technologies, and shared hopes for a better world.

They are fragile. They are complex. But they are also opportunities—to build a global order that is more responsive, more inclusive, and more just.

Let this book be not the end of a conversation, but the beginning of a commitment. To rethink diplomacy not just as negotiation, but as stewardship. To use technology not merely to expand reach, but to deepen trust. And to embrace this uncertain future not with fear—but with the conviction that the best diplomacy is always ahead of its time.

The future is digital.

But diplomacy must remain human.

Chapter 27: Epilogue

BEYOND THE SCREEN – THE LEGACY OF CONNECTION

The digital world is not a parallel universe. It is our world—amplified, accelerated, and interconnected more than ever before. As we close this exploration of diplomacy's transformation in the digital age, we do so with a simple but profound realization: the future is not a destination. It is something we are building, click by click, message by message, alliance by alliance.

Digital diplomacy is no longer niche. It is now core to how nations express their interests, defend their values, and respond to crises. It is how ideas gain traction, how trust is negotiated, how peace is proposed—and, at times, how it is fractured.

But technology alone cannot save us. Code is not conscience. Algorithms do not govern wisely by default. They inherit our intentions, our flaws, and our biases. That is why this era demands a new kind of diplomat: part engineer, part philosopher, part bridge-builder—anchored in values, but fluent in systems.

This book has not aimed to provide all the answers. Instead, it has tried to ask the right questions and illuminate the pathways being forged across continents and cloud networks alike. It is a chronicle of courage and caution, of progress and pitfalls. And it is a call to action for those who believe that diplomacy, at its best, can be an architecture of hope.

In this era of transformation, we must remember: the tools may change, but the purpose must remain—**to serve humanity, protect peace, and build a world where power is not only connected, but accountable.**

Our alliances may be wired.

But our diplomacy must always remain human.

Glossary of Digital Diplomacy Concepts

1. Digital Diplomacy

The use of digital technologies, platforms, and tools by states, diplomats, and international organizations to achieve foreign policy goals, engage with publics, and conduct negotiations.

2. Cyber Diplomacy

Diplomatic engagement focused on cybersecurity, cyber norms, and governance of cyberspace, often involving cross-border coordination on threats, responses, and legal frameworks.

3. E-Government

The digital delivery of government services and information through online platforms, aimed at improving efficiency, accessibility, and transparency.

4. Digital Sovereignty

The capacity of a state to independently manage, control, and protect its digital infrastructure, data, and policies without reliance on foreign entities.

5. Multilateral Digital Governance

Collaborative, international efforts to create common norms, rules, and standards for the digital realm across borders and institutions.

6. Digital Identity

An individual's representation in digital systems, often verified through biometrics, blockchain credentials, or national ID databases, granting access to services and legal recognition.

7. Blockchain Diplomacy

The use or regulation of blockchain technologies in diplomatic, governance, and development contexts, including smart contracts, decentralized organizations, and transparency systems.

8. Information Warfare

Strategic dissemination or manipulation of information—often via digital platforms—for political, military, or ideological influence during conflict or competition.

9. Digital Human Rights

The extension of traditional human rights to digital environments, including data privacy, freedom of expression, and protection from online harm.

10. Artificial Intelligence in Diplomacy

The application of AI technologies to analyze, predict, or assist in diplomatic decision-making, as well as the development of AI-related global governance standards.

11. Crisis Communication (Digital)

The real-time management of public messaging and information flow during crises via digital platforms, often involving governments, embassies, and international responders.

12. Data Localization

Policies requiring that data collected within a country's borders be stored and processed domestically, often tied to concerns about sovereignty and security.

13. Splinternet

The fragmentation of the internet into national or ideological silos, where different parts of the world operate under incompatible digital norms, censorship policies, or access restrictions.

14. DAOs
(Decentralized Autonomous Organizations)
Digital organizations governed by smart contracts and community consensus, without centralized leadership—explored for their potential role in decentralized governance and activism.

15. Techno-Authoritarianism
The use of advanced technologies by governments to exert control over citizens, suppress dissent, and monitor or manipulate behavior, often with limited oversight or recourse.

16. Digital Inclusion
Efforts to ensure that all individuals and communities, especially marginalized or underrepresented groups, have equitable access to digital technologies and skills.

17. Hybrid Threats
A blend of cyber, digital, and traditional geopolitical tactics used by state or non-state actors to destabilize, mislead, or coerce other nations or populations.

18. Norms of Responsible State Behavior in Cyberspace

Voluntary guidelines and principles negotiated through international forums to reduce the risk of cyber conflict and promote peaceful state conduct online.

19. Internet Governance

The distributed and multi-stakeholder process of shaping the evolution, use, and regulation of the internet, involving governments, corporations, civil society, and academia.

20. Smart Power (Digital)

The strategic use of digital tools and platforms to blend soft power (influence through culture and values) with hard power (economic or security leverage) in foreign policy.

Bibliography

This comprehensive bibliography provides a diverse range of sources that underpin the themes, discussions, and case studies presented throughout **"Diplomacy Unplugged: Reimagining Global Engagement in the Digital Age."** The selected works encompass books, academic articles, policy papers, reports, and credible online resources that enrich the understanding of the interplay between digital technologies, diplomacy, and international relations.

BOOKS & ACADEMIC SOURCES

Nye, Joseph S. *The Future of Power.* PublicAffairs, 2011.

Slaughter, Anne-Marie. *The Chessboard and the Web: Strategies of Connection in a Networked World.* Yale University Press, 2017.

Morozov, Evgeny. *The Net Delusion: The Dark Side of Internet Freedom.* PublicAffairs, 2011.

Mueller, Milton. *Networks and States: The Global Politics of Internet Governance.* MIT Press, 2010.

Deibert, Ronald. Reset: *Reclaiming the Internet for Civil Society*. House of Anansi Press, 2020.

West, Darrell M. *The Future of Work: Robots, AI, and Automation*. Brookings Institution Press, 2018.

Donahoe, Eileen, and Megan MacDuffee Metzger. "Promoting Human Rights in the Age of AI." *Journal of Democracy*, vol. 31, no. 2, 2020.

REPORTS & INSTITUTIONAL PUBLICATIONS

United Nations. *Roadmap for Digital Cooperation*. UN Secretary-General's High-level Panel on Digital Cooperation, 2020.

World Economic Forum. *Global Risks Report 2023*. Geneva: WEF, 2023.

OECD. *Digital Government Index 2022*. Organisation for Economic Co-operation and Development.

World Bank. *World Development Report 2021: Data for Better Lives*. Washington, DC: World Bank Group.

International Telecommunication Union. *Measuring Digital Development: Facts and Figures 2023*. Geneva: ITU, 2023.

UNESCO. *Internet Universality Indicators: A Framework for Assessing Internet Development*. Paris: UNESCO, 2020.

European Commission. *Artificial Intelligence Act Proposal. Brussels:* European Union, 2021.

UNDP. Digital Strategy 2022–2025: *Building the Digital Commons.* New York: United Nations Development Programme.

ARTICLES & POLICY BRIEFS

Pamment, James. "Digital Diplomacy as Transversal Practice." *The Hague Journal of Diplomacy*, vol. 11, no. 1, 2016.

Bjola, Corneliu, and Marcus Holmes, eds. *Digital Diplomacy: Theory and Practice*. Routledge, 2015.

Kurbalija, Jovan. *An Introduction to Internet Governance.* DiploFoundation, 7th edition, 2020.

Zwitter, Andrej, and Oskar J. Gstrein. "Big Data, Privacy and COVID-19 – Learning from Humanitarian Expertise in Data Protection." *International Journal of Humanitarian Action*, 2020.

CASE STUDY SOURCES

Bellingcat. *Investigative Reports and Methodologies.* www.bellingcat.com

UNHCR. *Digital Identity for Refugees: UNHCR Innovation Service.* www.unhcr.org

Estonia e-Residency Program. *White Papers and Government Reports.* www.e-resident.gov.ee

Aadhaar Project (UIDAI, India). *Annual Reports and Policy Documents.* www.uidai.gov.in

UAE Smart Government Strategy. *Ministry of Cabinet Affairs and The Future.* www.smartdubai.ae

NEWS & MEDIA COVERAGE

The Economist. Special Report on Digital Diplomacy. 2022.

Foreign Affairs. "The Techno-Geopolitical Arms Race." March/April 2023.

The Guardian. "Facebook's Role in Myanmar Genocide." 2018.

Al Jazeera. "Climate Diplomacy and the Youth Movement." 2019.

Reuters. "G7 Nations Agree on AI and Cybersecurity Standards." 2023.

ONLINE DATABASES AND RESEARCH PORTALS

1. Google Scholar: A specialized search engine providing access to scholarly articles, theses, books, and conference papers impacting the field of digital diplomacy. **Website:** Google Scholar

2. JSTOR: An expansive digital library containing academic journal articles, books, and primary sources related to international relations and diplomacy. **Website:** JSTOR

3. ResearchGate: A networking site for researchers to share their findings, collaborate, and engage with the academic community. The platform often includes access to papers and articles relevant to digital diplomacy. **Website:** ResearchGate

Made in the USA
Columbia, SC
08 August 2025

d1c27c08-6f49-4e84-9d47-57b2eb27fe50R02